**Microgrid Planning and Design**

# Microgrid Planning and Design

A Concise Guide

*Hassan Farhangi*
British Columbia Institute of Technology
Canada

*Geza Joos*
McGill University
Canada

*Registered Offices*
John Wiley & Sons, Inc., 111 River Street, Hoboken, NJ 07030, USA
John Wiley & Sons Ltd, The Atrium, Southern Gate, Chichester, West Sussex, PO19 8SQ, UK

*Editorial Office*
The Atrium, Southern Gate, Chichester, West Sussex, PO19 8SQ, UK

For details of our global editorial offices, customer services, and more information about Wiley products visit us at www.wiley.com.

Wiley also publishes its books in a variety of electronic formats and by print-on-demand. Some content that appears in standard print versions of this book may not be available in other formats.

*Library of Congress Cataloging-in-Publication data applied for*

ISBN: 9781119453505

Cover Design: Wiley
Cover Image: Courtesy of British Columbia Institute of Technology (BCIT)

Set in 10/12pt WarnockPro by SPi Global, Chennai, India
Printed in Singapore by C.O.S. Printers Pte Ltd

10  9  8  7  6  5  4  3  2  1

Knock, *And He'll open the Door*
Vanish, *And He'll make you shine like the Sun*
Fall, *And He'll raise you to the Heavens*
Become nothing, *And He'll turn you into Everything.*

~Rumi

Dedicated to practitioners who embark on the long journey of discovery for the betterment of mankind, knowing that the road ahead is often paved with quandary and enigma, rather than triumph. Nevertheless, what defines such a voyage is that first step, taken in the hope that each subsequent one would bring humanity a little closer to the desirable future imagined today.

This book attempts to narrate one such journey, and its many successes and failures.

…….and to Mohammad-Ali Samimi & Sareh Farhangi who were the true beacons of my life!

Hassan Farhangi

# Contents

# About the Authors

**Hassan Farhangi, PhD, SMIEEE, PEng**, is the director of smart grid research at British Columbia Institute of Technology (BCIT) in Burnaby, British Columbia, Canada, and adjunct professor at Simon Fraser University in Vancouver, Canada. Dr. Farhangi has held adjunct professor appointments at the National University of Singapore, Royal Road University, in Victoria, Canada, and at the University of British Colombia, in Vancouver, Canada. Dr. Farhangi is currently the chief system architect and the principal investigator of BCIT's smart Microgrid initiative at its Burnaby campus in Vancouver, British Columbia, and the scientific director and principal investigator of Natural Sciences and Engineering Research Council's (NSERC) Pan-Canadian smart Microgrid network (NSERC Smart Microgrid Network or NSMG-Net). He has published widely, with numerous contributions in scientific journals and conferences on smart grids and has served on various international standardization committees, such as International Electrotechnical Commission (IEC) Canadian Subcommittee (CSC) Technical Committee 57 (TC 57) Working Group 17 (WG 57) (IEC 61850), Conseil International des Grands Réseaux Électriques (CIGRÉ) WG C6.21 (Smart Metering), CIGRÉ WG C6.22 (Microgrids Evolution), and CIGRÉ WG C6.28 (Hybrid Systems for Off-Grid Power Supply). Dr. Farhangi obtained his PhD degree from the University of Manchester Institute of Science and Technology, in the United Kingdom, in 1982; his MSc degree from the University of Bradford, in the United Kingdom, in 1978; and his BSc degree from the University of Tabriz, in Iran, in 1976, all in electrical and electronic engineering. Dr. Farhangi is a founding member of SmartGrid Canada, an academic member of CIGRÉ, a member of the Association of Professional Engineers and Geoscientists of British Columbia, and a senior member of the Institute of Electrical and Electronic Engineers.

**Geza Joos, PhD, FIEEE** is a Professor in the Department of Electrical and Computer Engineering, McGill University, and holds the NSERC/Hydro-Quebec Industrial Research Chair on the Integration of Renewable Energies and Distributed Generation into the Electric Distribution Grid and the Canada Research Chair in Powering Information Technologies (Tier 1) at McGill University. He has also been involved in industrial consulting and in industry R&D management as technology coordinator for the Power System Planning and Operation Interest Group at CEATI International. His expertise is in power electronics, with applications to power systems and energy conversion. Recent topics have dealt with integration of renewable energy, mainly wind, and distributed generation. He has published numerous journal and conference papers and presented tutorials at international conferences on these subjects. He is active in several Institute of Electrical and Electronic Engineers (IEEE) Power Engineering Society and CIGRE working groups dealing with these issues. He is a Fellow of the IEEE and an active researcher and collaborator with other team members of the NSMG-Net Strategic Research Network led by Dr. Farhangi and hosted at BCIT. Dr. Joos has supervised many Masters and PhD students at McGill.

# Disclaimer

This book was prepared as the result of research conducted by the NSERC Strategic Research Network in Smart Microgrids (NSMG-Net), comprised of researchers from various universities and academic institutions from across Canada. Nevertheless, this book does not necessarily represent the views of these individuals, their universities, their partners, or the funding agencies that have funded their work. As such, the authors make no warrant, express or implied, and assume no legal liability for the information in this book; nor does any party represent that the uses of this information will not infringe upon privately owned rights. This book has not been approved or disapproved by the authors, nor have they certified the accuracy or adequacy of the information in this book. Moreover, the information contained in this book is subject to future revisions. Important notes of limitations include the following: this book is not a replacement for electrical codes or other applicable standards; this book is not intended or provided by the authors as a design specification or as design guidelines for electrical installations; and the book shall not be used for any purpose other than education and training. Persons using this information do so at no risk to the authors, and they rely solely upon themselves to ensure that their use of all or part of this book is appropriate in the particular circumstance.

## List of Figures

# List of Tables

# Foreword

The NSERC Smart Microgrid Network (NSMG-Net) was launched in late 2010 to address a critical and growing need in the electricity industry to transform its existing, outdated power grid into a next generation intelligent (a.k.a. smart) grid. The primary building blocks of a smart grid are smart microgrids, which are geographically compact units with a flexible distribution system integrated with the main power grid, which can be connected and disconnected to run autonomously for self-sufficiency of energy production and consumption. By definition, each smart microgrid is capable of engaging in energy transactions with other microgrids as well as with a central utility command and control infrastructure, and has geographical attributes and functions that apply to the local area in which it operates.

At the time NSMG-Net was initiated, the technologies required to develop smart microgrids were in the research phase. Full development of smart microgrids in the utility context required research with respect to each jurisdiction's specific climate, terrain, and available energy sources, as well as testing, verification, and qualification in near-real environments. The overarching goal of NSMG-Net was thus to develop the building blocks of a new smart electrical microgrid that could ultimately provide reliable, low cost, and clean power to communities across the globe.

NSMG-Net started its research in 2010, comprising researchers from universities of New Brunswick, McGill, Toronto, Ryerson, Waterloo, Manitoba, Alberta, Simon Fraser, British Colombia (UBC) and hosted and led by British Columbia Institute of Technology (BCIT). Adopting an interdisciplinary research strategy, and capitalizing on BCIT's Smart Microgrid testbed, NSMG-Net researchers achieved significant progress in the development of technologies, know-how and models for smart microgrid system deployment over the six years of their focused research. Considered as one of the most successful Strategic Research networks, funded by NSERC to date, NSMG-Net published more than 80 papers in peer-reviewed journals, and made numerous conference presentations in major academic and scientific gatherings across the world. The Network also trained over 130 engineering students, including 37 undergraduates, 50 masters, 48 doctorate, 5 post-doctoral fellows and 3 research associates. Four patent applications were filed, together with eight IP disclosures. More importantly, there was significant engagement with several industry partners and other stakeholders both nationally and internationally.

*Hassan Farhangi,*
NSMG-Net Principal Investigator and
Scientific Director

# Preface

This book provides a microgrid design guidelines framework, consisting of specification requirements, design criteria, recommendations, and sample applications for the stakeholder/client in order to comprehend the technologies, limits, tradeoffs, and potential costs and benefits of implementing a smart microgrid. The guidelines take into account such diverse applications of microgrids as required by urban, mining, campus, and remote communities. The development of these guidelines was carried out using a systematic design methodology that comprises design criteria, modeling, simulations, economic and technical feasibility studies, and business case analysis. In addition, the guidelines address the real-time operation of the microgrid (voltage and frequency control, islanding, and reconnection) as well as the energy management system in islanded and grid-connected modes. It also summarizes available microgrid benchmarks for the electric power system, control systems, implementation approaches, and the information and communication systems implementable in microgrids. It covers a modeling approach that combines the power and communication systems, allowing a complete system study to be conducted. The book includes the development of use cases and the validation of the models with field results from the BCIT Microgrid and the IREQ test line.

This book is essentially a compilation of research work performed by NSMG-Net researchers, and published in the public domain, between 2010 and 2016.

*Geza Joos*
NSMG-Net Theme 2 Leader and
Chair Outreach Committee

# Acknowledgments

This book would not have been possible without the efforts and contributions of the NSERC Smart Microgrid Network (NSMG-Net) researchers, students, associates, partners, and funders. In particular, the efforts by the NSMG-Net theme leaders, topic leaders, network researchers, students, and supporting staff during the six years of the network's research are acknowledged. This book capitalizes on numerous papers, reports, and documentations issued by network researchers, including interim reports, annual reports, publications, theses and dissertations, workshops and training materials. The design guidelines, discussed in the body of this book, were extracted, assembled, and inferred out of numerous works published by NSMG-net researchers. Furthermore, the assistance provided by researchers at the British Columbia Institute of Technology (BCIT), Burnaby, BC and the Institut de recherche d'Hydro-Québec (IREQ), Varennes, QC is also acknowledged. A special gratitude to our industry and utility partners whose contribution in stimulating suggestions and encouragement, helped in writing these important design guidelines. Finally, the authors would like to acknowledge the information and publications provided by the following researchers, and their students who participated in the NSMG network, including, but not limited to, and in no particular order: Dr. Reza Iravani, University of Toronto, Ontario, Canada; Dr. Geza Joos, McGill University, Quebec, Canada; Dr. Fabrice Labeau, McGill University, Quebec, Canada; Dr. Tho Le-Ngoc, McGill University, Quebec, Canada; Dr. Dave Michelson, University of British Columbia, BC, Canada, Dr. Ani Gole University of Manitoba, Manitoba, Canada; Dr. Wilsun Xu, University of Alberta, Alberta, Canada; Dr. Julian Meng, University of New Brunswick, NB, Canada, Dr. Eduardo Castillo Guerra, University of New Brunswick, NB, Canada; Dr. Kankar Bhattacharya, University of Waterloo, Ontario, Canada; Dr. Amirnaser Yazdani, Ryerson University, Ontario, Canada; Dr. Hassan Farhangi, BC Institute of Technology, Vancouver, BC, Canada; Dr. Ali Palizban, BC Institute of Technology, Vancouver, BC, Canada; Dr. Siamak Arazanpour, Simon Fraser University, Vancouver, BC, Canada; Dr. Mehrdad Moallem, Simon Fraser University, Vancouver, BC, Canada; Dr. Gary Wang, Simon Fraser University, Vancouver, BC, Canada; Dr. Daniel Lee, Simon Fraser University, Vancouver, BC, Canada.

## Acronyms and Abbreviations

| | |
|---|---|
| AMI | Advanced Metering Infrastructure |
| Area-EPS | Area Electric Power System |
| BCIT | British Columbia Institute of Technology |
| BESS | Battery Energy Storage System |
| C | Controlled (Loads) |
| CB | Circuit Breaker |
| CEC | Canadian Electrical Code |
| CHIL | Controller Hardware in the Loop |
| CHP | Combined Heat and Power |
| CSA | Canadian Standards Association |
| DEMS | Decentralized Energy Management System |
| DER | Distributed Energy Resource |
| DER-CAM | Distributed Energy Resources Customer Adoption Model |
| DERMS | Distributed Energy Resource Management System |
| DG | Distributed Generation |
| DMS | Distributed Management System |
| DR | Demand Response |
| DSM | Demand Side Management (also response or integration) |
| DSO | Distribution System Operator |
| DSP | Digital Signal Processing |
| EC | Enhanced Control |
| EMS | Energy Management System |
| EPRI | Electric Power Research Institute |
| EPS | Electric Power System |
| ESS | Electrical Storage System, or Energy Storage System |
| EV | Electric Vehicle |
| EVFC | Electrical Vehicle Fast Chargers |
| FAT | Factory Acceptance Test |
| FERC | Federal Energy Regulatory Commission |
| FIT | Factory Integration Test |
| GHG | Green House Gases |
| GOOSE | Generic Object Oriented Substation Event |
| GPS | Global Positioning System |
| GSP | Grid Supply Point |
| GW | Gigawatts - unit of power |

| | |
|---|---|
| HIL | Hardware in the Loop |
| HMI | Human Machine Interface |
| HV | High Voltage |
| ICT | Information and communication technology |
| IEC | International Electrotechnical Commission |
| IEC 61850 | IEC # 61850 – Communication networks and systems in substations |
| IEEE | Institute of Electrical and Electronics Engineers |
| IEEE 519 | IEEE Recommended Practices and Requirements for Harmonic Control in Electrical Power Systems |
| IEEE 1547 | IEEE Standard for Interconnecting Distributed Resources with Electric Power Systems |
| ISO | Independent System Operator |
| Local-EPS | Local Electric Power System |
| LPSP | Loss of Power Supply Probability |
| LVRT | Low Voltage Ride Through |
| MCC | Microgrid Central Controller |
| NERC | North American Electric Reliability Corporation |
| NSERC | Natural Sciences and Engineering Research Council |
| NSMG-NET | NSERC Smart Microgrid Network |
| NSMN | NSERC Smart Microgrid Network |
| OASIS | Open Access to Sustainable Intermittent Sources |
| PCC | Point of Common Coupling |
| P-f | Active power and frequency control |
| PHIL | Power Hardware in the Loop |
| PMU | Phase Measurement Unit |
| P-Q | Active and Reactive power control |
| PV | Solar Photovoltaic |
| RES | Renewable Energy Sources |
| RTS | Real Time Simulation/Simulator |
| SCADA | Supervisory Control and Data Acquisition |
| SG | Synchronous Generator |
| UF | Under-frequency |
| UML | Unified Modeling Language |
| V-P | Voltage and Active power control |
| V-Q | Voltage and Reactive power control |
| U/C | Uncontrolled (Loads) |
| UV | Under-voltage |

# 1

## Introduction

In the face of rising demand for electricity amid increasing costs and environmental impacts related to burning fossil fuels, utility companies must research ways to manage demand and integrate renewable sources of energy into the mainstream power system. North American utilities have begun setting targets to meet some of their future growth in electricity demand through energy conservation. Many utility companies have devised elaborate technology roadmaps and implementation plans to rejuvenate their aging infrastructure. The main barrier to moving forward for these and other power suppliers is the current antiquated nature of the electricity grid. Designed to cater to a centralized power generation, transmission, and distribution system, it does not readily lend itself to accommodating new technologies and solutions. It is the last remaining sector providing a critical service to customers without having real-time feedback about how its services are utilized by its users. The electricity grid has traditionally operated as an open-loop system where no real-time data have been captured for such things as instantaneous demand, consumption profiles, or system performance. This open-loop system cannot store energy, nor can it integrate renewable sources of energy, such as wind, solar, biomass, and wave/tide, with their intermittent behaviors or embrace pervasive control systems required to attain operational efficiencies and energy conservation.

Built in the last century, the electricity grid is a one-way hierarchical system whereby power is generated and dispatched based on historical consumption data rather than on real-time demand. As such, the system is over-engineered by design to withstand peak loads, which might not be present at all times. This means that the system's expensive assets are not efficiently used at all times. Overall system control is achieved through an elaborate frequency regulation scheme. The control leverage is largely exercised at the production end through statistically planned responses to anticipated changes in system frequency. When the demand increases, the drop in system frequency is countered by increasing the system's production through leveraging spinning reserves until system frequency is back to normal. In contrast, when the demand decreases, resulting in a sudden rise in system frequency, the system responds by decreasing system production until system frequency is reduced to nominal values. However, this synchronous control is slow in nature and if variations in the forecasted demand happen faster than the system's intrinsic inertia can deal with, the system simply fails, resulting in brownouts and blackouts.

Electricity generation, transmission, and distribution are supported by an electricity grid, which forms the backbone of a typical power network. A grid may reference a sub-network, such as a local utility's transmission grid or distribution grid, a regional

*Microgrid Planning and Design: A Concise Guide,* First Edition. Hassan Farhangi and Geza Joos.
© 2019 John Wiley & Sons Ltd. Published 2019 by John Wiley & Sons Ltd.

transmission network, a whole country's or entire continent's electrical network. Conventional centralized grid systems tend to experience significant power loss due to inefficiencies. They are also being challenged by increasing demand, rising costs, tightening supply, declining reserve margins, and the need to minimize environmental impacts. At no time in its century-long history has the global utility industry had to confront so many diverse and concurrent challenges as it does now. In the last few decades, electrical power providers have faced one or more of the following challenges:

- Aging infrastructure (more than 70% of utility assets in the USA are over 25 years old);
- Reliability (rampant blackouts in California, Northeast USA, and in Eastern Canada);
- Security (researchers in the USA have proven that the US electrical grid is prone to attacks);
- Market dynamics (various jurisdictions are moving toward industry deregulation);
- Rates and pricing (need to implement multi-tariffs, time of use, smart metering, etc.);
- Distributed generation (DG) (the need to allow access to the grid by Independent Power Producers (IPPs) and Co-Gens);
- Efficiency and optimization (need for demand response and peak control);
- Rising energy costs (related to rising oil prices and security of supply);
- Conservation (of the planet's limited source of energy);
- Mass electrification (meeting increasing demands on electricity);
- Renewable energy (integration of renewable sources of energy into the grid); and
- Green energy (minimizing the industry's carbon footprint).

At the core of the crisis is the inability of conventional electrical grids to respond to such challenges without major technological overhaul of their infrastructure. Among other things, this overhaul requires a layer of intelligent command and control to be placed on top of the electricity grid. Unfortunately, this level of intelligence cannot be introduced within the framework of utilities' existing electricity grids. A new and improved electrical grid is required. How this new grid, known as either an 'intelligent grid' or a 'smart grid' – in this book, 'intelligent grid' and 'smart grid', as well as 'intelligent microgrid' and 'smart microgrid' will be used interchangeably – differs from the current model as discussed here.

As Table 1.1 depicts, the next generation grid, is a convergence of information technology and communication technology with power system engineering. Smart grid is the focus of assorted technological innovations, which utility companies throughout North America and across the world plan to incorporate in many aspects of their operations and infrastructure. Given the sheer size of utility assets, the emergence of smart grid is more likely to follow an evolutionary trajectory rather than a drastic overhaul.

Smart grid will therefore materialize through strategic implants of distributed control and monitoring systems within and alongside existing electricity grids. Smart grids' functional and technological growth will mean that pockets of distributed intelligent systems emerge across diverse geographies. This organic growth will allow the utility industry to shift more of the old grid's load and functions onto the new grid, thus improving and enhancing their critical services. These smart grid embryos, known as intelligent or smart microgrids, will facilitate DG and co-generation of energy. They will also provide for the integration of alternative sources of energy and management of the system's emissions and carbon footprint. They will enable utilities to make more efficient use of their existing assets through demand response, peak shaving, and service quality control.

**Table 1.1** Smart grid vis-à-vis the existing grid.

| Existing grid | Smart grid |
|---|---|
| Electromechanical | Digital |
| One-way communication | Two-way communication |
| Centralized generation | Distributed generation |
| Hierarchical | Network |
| Few sensors | Sensors throughout |
| Blind | Self-monitoring |
| Manual restoration | Self-healing |
| Failures and blackouts | Adaptive and islanding |
| Manual check/test | Remote check/test |
| Limited control | Pervasive control |
| Few customer choices | Many customer choices |

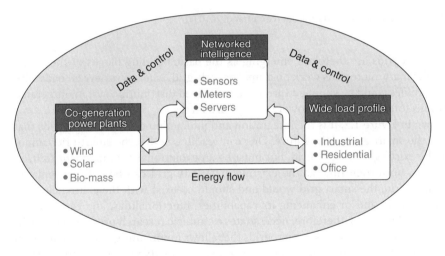

**Figure 1.1** Topology of a smart microgrid.

However, the problem that most utility providers across the globe face is how to get to where they need to be as soon as possible, at the minimum cost, and without jeopardizing the critical services they are currently providing. Moreover, power companies must decide what strategies and along what pathways they should choose to ensure the highest possible return on the required investments for such major undertakings. At its core, the smart grid may emerge as an ad hoc integration of complementary components, subsystems, and functions under the pervasive control of a highly intelligent and distributed command and control system, developed by assimilating smart microgrids. As Figure 1.1 depicts, intelligent or smart microgrids form an interconnected network of distributed energy systems (loads and resources) that can function connected to, or separate from the overall electricity grid.

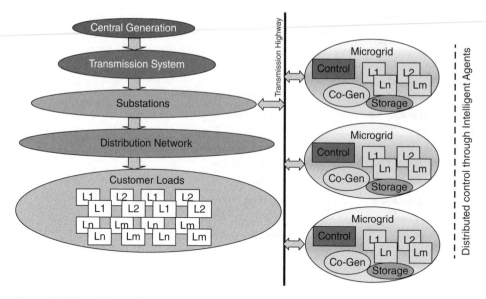

**Figure 1.2** The evolution of smart grid. Source Farhangi 2010 [1].

Smart microgrids are therefore emerging as the basic building blocks of the future smart grid. Smart microgrids work on a smaller scale grid, where a variety of loads with different profiles could be supplied through a controlled distribution system integrated with various (often renewable) power generation sources.

As shown in Figure 1.2, it is the integration and interaction of smart microgrids that will form the smart grid of the future. One can readily see that the smart grid cannot replace the existing electricity grid, but must be an evolutionary complement to it. In other words, until such time as the existing electricity grid can be entirely replaced by the smart grid, the smart grid would and should co-exist with the exiting electricity grid, adding to and/or enhancing its capabilities, functionalities, and capacities as it further develops. This, therefore, necessitates a strategic research program into technologies and standards for smart microgrids that allows for organic growth, inclusion of forward-looking technologies and full backward compatibility with the existing legacy systems. The research program should therefore focus on three complementary areas:

- Operational issues of a smart microgrid, such as protection, switching, dispatching, control, management, etc.;
- Regulatory and standardization of the components, interfaces, and subsystems of a smart microgrid; and
- Communication, messaging, data networking, and automation of smart microgrid components and systems.

One major hurdle that has prevented utility companies from venturing into smart grid development is the need to test and validate technologies and solutions within a near-real environment before such sub-systems could be regarded as grid-worthy. Given the fact that power companies must constantly provide a service critical to society, it is logical that successful lab tests or small-scale pilots would not qualify such new technologies to be candidates for integration into such crucial infrastructure as the electricity grid. For that reason, there is a need for a microgrid environment where smart grid technologies at a sufficiently large scale could be developed, integrated as desired solutions, tested, and qualified. In many parts of the world, attempts are underway to set up microgrid infrastructure for research and development. As scaled-down versions of the intelligent grid, smart microgrids enable utility companies, technology providers, and researchers to work together to develop and facilitate the commercialization of architectures, protocols, configurations, and models of the evolving intelligent grid with the intention of charting a "path from lab to field" for innovative and cost-effective technologies and solutions for the evolving smart electricity grid.

Such smart microgrids are typically designed as an RD&D (Research, Development, and Demonstration) platform where existing and future technologies in telecommunication, smart metering, co-generation, and analytics are employed to develop and qualify the most robust, cost-effective and scalable solutions required to facilitate and nurture the evolution and the emergence of the smart grid in one form or another.

Such initiatives are best served through consortiums of public and private players. A consortium, composed of technology companies at the forefront of technology development in the field, as well as utility companies representing the users, supported by funding agencies, can focus on the development of a wide array of hardware, software, and system technologies required for the realization of an intelligent grid. The validation and qualification of architectures, models, and protocols developed are typically guided, supervised, and scrutinized by partnering utility companies who act as the end-customers of the developed technologies and solutions. This so-called RD&D network allows all the stakeholders of a region or country to come together to develop solutions that cater to that specific region, country, or geographic area.

## 1.1 Why Microgrid Research Requires a Network Approach

The need for a national (or regional) network in any country (or region) for research in developing a smart grid is mainly based on two distinct conditions:

- The national electricity grid in any country has evolved over the last century based on that country's specific economy, geography, and climate. While generation capacity in certain countries may have been dominated by hydroelectric power, other countries may have used nuclear or fossil fuel-based energy generation, while still others may have attempted to tap into renewable sources of energy. The transmission side of the utility industry has also seen its own share of peculiarities, reflecting each country's geographical and climate conditions. Moreover, different urbanization patterns in each country may have influenced the distribution system design. Developing the next generation and future smart grid for any country requires attention to be paid to the same realities and patterns of its existing electrical grid while responding to the challenges of providing power to the country as a whole, and so must involve researchers and utility companies from across the country. Many countries in the world have already started major research initiatives in this area, with the understanding that with very little exception each country will have to develop their own specific blueprint for the design of their national smart grid.

- The smart grid is regarded as the true convergence of Information Technology, Communication Technology, and Power Systems. It is viewed as a collection of concepts, technologies, and approaches to how energy is best harnessed, exploited, and utilized. This means that the development of such advanced concepts requires active integration of technology developers, academia, end-customers, and regulatory authorities from across the country. The network approach is thus critical to building the foundation of the next-generation electric power grid for all countries. Such national research networks will seek and encourage proactive participation by many researchers, utilities, and technology providers from across the country in helping them achieve the goals of such national effort.

Shifting from the current centralized electricity grid model to a distributed smart grid model requires more than merely implementing new technologies. It also requires a significant cultural, philosophical, and technological change in the exploitation of energy sources. In the twentieth century, electricity production was centralized, under the control of single provincial entities and focused on providing large amounts of power from few sources. In the twenty-first century this model has shifted to one focused on integrating multiple clean sources of energy in a flexible secure manner that can meet customers' changing needs over time. Individual customers, governments, and industry have all signaled that they are ready to make this shift. However, there are barriers to smart grid implementation that need to be overcome.

For example, numerous research topics are being conducted on renewable sources of energy, but even if successful solutions are found, few of them will ever find their way into the mainstream power system because of a lack of cost-effective technologies, interface standards, application protocols, integration systems, reliable storage, appropriate policies, and efficient transmission/distribution infrastructure. The multifaceted nature of the issues and the need to leverage the expertise of a multidisciplinary, multi-sector group of experts necessitates such research to be conducted through a well-organized and closely coordinated network approach.

A strong feature of network research model is the participation of utility companies as sponsors and partners, which are in a position to not only validate network approaches and directions, but also adopt and utilize research findings and developments. In addition to utility companies, high-tech partners have the capability and interest to further develop the Network's research outcomes and take these to market as commercial products. Such close cooperation shall ensure the viability and feasibility of research outcomes. Research into smart microgrids requires systemic collaboration between scientists, end customers, and commercialization partners within a multidisciplinary research program to overcome the above impediments and address the strategic need to ensure reliable, cost-effective and secure electricity supply for generations to come.

Smart microgrids can be viewed as a first step toward the implementation of smart grids, on a smaller and more manageable scale. Smart grid technologies are a series of innovations that combine technologies incorporating such diverse fields as power systems, power electronics and control, communications and measurements, and intelligent sensors and metering. The development and deployment of these technologies will require a multidisciplinary research effort, which could only be met through clusters of researchers. It will bring together experts in power electronics and control, power systems operation, energy management and modeling, and communication protocols, sensors and condition monitoring, and integrated data management.

As an example of the complementary nature of the research, integrating renewable sources energy, due to their variable and intermittent nature, requires the following approaches: (i) control of the renewable energy resource, including storage; (ii) management of the energy produced and exploitation of the ancillary services that can be provided; (iii) sensing, monitoring, communications and data management to ensure proper integration and control of the renewable resource. The network approach allows researchers from diverse fields to work together to develop solutions that will facilitate, in the example cited earlier, the integration of renewable energy into distribution grids. Such varied expertise, not found in any one given university in any jurisdiction, would provide the critical mass required to carry out advanced research in intelligent grids.

Networked research allows for collaboration between researchers across many jurisdictions who have access to electrical utilities and industries that work in different contexts and with differing constraints. This widens the scope of issues that can be dealt with in the context of smart microgrid deployment and allows solutions to cover a wider spectrum of issues. In addition to bringing together researchers in diverse fields, the Network will allow students from different institutions an opportunity to receive training in a combination of technical fields, and exposure to the operation of power systems and to new initiatives developed by utility companies. These companies will be looking increasingly for experts with a multidisciplinary background. One of the more important combinations being sought at this time is a basic training in power systems with an advanced expertise in communications. A research network is ideally suited to train such highly qualified personnel.

Canada's first smart microgrid infrastructure began to take shape in 2008 at BCIT's Main Campus in Burnaby. It enabled high-tech companies, end customers, and researchers to work together to develop and qualify various system architectures, configurations, interface protocols, and grid designs to meet national and global priorities for co-generation, efficient dispatch and distribution of electricity, load control, demand response, advanced metering and integration of clean energy sources into the existing and future grids.

## 1.2 NSERC Smart MicroGrid Network (NSMG-Net) – The Canadian Experience

Upon completion of BCIT's smart microgrid in 2010, the Natural Sciences and Engineering Research Council of Canada (NSERC) funded a pan-Canadian Strategic Microgrid Research Network (called NSERC Smart Microgrid Network [NSMG-Net]) to bring together world-class researchers, from many prestigious universities across Canada, to utilize BCIT's smart microgrid infrastructure for test and validation of their research results. NSMG-Net, acted as the platform needed to support integration and testing activities, enabling researchers to pool their diverse expertise and resources from across the country to develop smart grid technologies and solutions to be customized for the different climate, geographic, and economic demands found across the globe. The present book attempts to compile and document the results of the research program that the network set out to conduct with an underlying research question of *what are the design guidelines for smart microgrids?*

Upon its inception, as a pan-Canadian strategic research network, NSMG-NET took upon itself the critical task of bringing together some of Canada's best researchers and resources in the then relatively unknown smart grid and microgrid areas to focus on the development of technologies and know-how required to build a scaled-down version of smart grid, namely a smart microgrid. The challenge was to create an environment in which researchers with different and divergent backgrounds, and no history of collaborative work, would find a common vocabulary to communicate, exchange information and find interdisciplinary solutions for the multifaceted problems facing our utility industry. The Network leaders knew from the start that bringing Power System Engineers, Communication System Engineers, and Information Technology Experts to work together was an exceptional challenge. Nevertheless, through careful planning and execution, the right frameworks and tools were put in place to ensure cross-pollination of ideas and know-how across various domains and boundaries in the Network.

The aim of these collaborations is to provide a guide to the form of technologies and solutions applicable to the design and/or operation of operational microgrids. The effort provided NSMG-Net researchers and particularly students, with the participation of industrial partners, a unique opportunity to apply the technologies they developed in a real-world microgrid, and test and verify the validity of their design approaches and solutions. It should be noted that many of the projects involved the participation, in one form or another, of Network partners from government, namely Natural Resources Canada's Canmet-Energy Laboratory, utilities, including BC Hydro, Hydro-Quebec, and Hydro One, and a number of industries. Project leaders and their students had many interactions with these partners, who provided, among others, information about the structure and operating conditions of electric grids, including remote and isolated grids, a specific Canadian reality, and about applicable power and communication technologies. The network was funded by NSERC from 2010 to 2016. However, the collaborations that resulted from the network are still bearing fruits. This book provides guidelines for microgrid design from the network research experience.

## 1.3   Research Platform

Academic research focuses on technology development. However, there are very few infrastructures available to test and validate those technologies in real conditions. Utility companies are also extremely cautious about introducing new technologies, solutions, and approaches in their highly critical infrastructure without extensive testing and validation. Traditionally, utilities have deployed fully integrated end-to-end proprietary solutions and technologies. In contrast, smart grid has created opportunities for the technological contributions and innovations of smaller companies who focus on individual pieces of the puzzle. In the absence of standards, protocols, and established interfaces, technologies developed by these companies need to be integrated, tested, and qualified within an impartial and non-threatening environment with complementary technologies supplied by other players to form the end-to-end solutions that the utilities require. This environment is commonly regarded as a microgrid. The values that a smart microgrid provides for researchers in the field are typically:

- It provides researchers, academia, and high-tech companies with an impartial environment in which their technologies can be integrated with other complementary technologies from other suppliers to form solutions required by Utilities.
- It provides Utilities with a neutral environment in which they can interact, discuss, and communicate their specifications, needs, and requirements with the research and high-tech community.
- It provides Utilities, researchers, and high-tech companies with a near-real environment in which the developed solutions and technologies could be integrated, tested, and exhaustively qualified before they could be regarded as "grid-worthy."
- It provides regulatory authorities with an environment in which new national standards, protocols, and models can be developed, tested, and qualified.
- It provides students with appropriate trainings to equip them with the new skills and techniques that will be required by utilities to develop and roll-out smart grid.

Ideally, a microgrid used for such purposes should be equipped with its own feeds from the utility company, as well as its own local substations and distribution infrastructure and its own co-generation capabilities. In addition to the termination points, localized co-generation plants (Solar, Wind, and Thermal) should preferably be present to allow for the development of Command and Control models for co-generation and integration of such alternative energy sources into the future smart grid. The network of smart terminations and control components should also be equipped with smart meters, appliances fitted with smart controls, and all these components be fitted with communication modules to enable them to communicate with data aggregation units in substations.

Moreover, the microgrid should cater to a wide variety of different electricity consumption profiles ranging from heavy industrial machinery to office type consumption along with residential-type profiles.

## 1.4 Research Program and Scope

What hinders wider adoption of alternative sources of energy, such as wind, solar, biomass, and wave/tide, is not solely their diffuse nature or higher cost, but also the absence of suitable devices, interface standards, application protocols, integration systems, reliable storage, appropriate policies, and efficient transmission/distribution infrastructure. Numerous research topics are being conducted on alternative energy technologies, but even if successful solutions are found, few of them will ever find their way into the mainstream power system if the complexities of their integration, as well as interaction, with the main grid are not adequately addressed.

As discussed earlier, smart microgrids are believed to become the building blocks of a new electrical grid that will: (i) provide reliable, low cost and clean power to communities across the globe; (ii) defer investments in transmission and distribution systems; (iii) improve power quality and reduce electrical system losses; (iv) improve energy efficiency and enable conservation; and (v) contribute to the reduction of energy system's carbon footprints. This would be achieved by transforming the energy production and transmission system to one that integrates two-way communication, smart components, renewable sources of energy and intelligent control into a sustainable energy system, thus improving reliability, security, cleanliness of energy, efficiency, self-sufficiency, and affordability.

## 1.5 Research Themes in Smart Microgrids

Given the typical structures that smart microgrids will ultimately assume, research work in this area could very well follow the following three themes:

Theme 1: Operation, control, and protection of smart microgrids;
Theme 2: Smart microgrid planning, optimization, and regulatory issues; and
Theme 3: Smart microgrid communication and information technologies.

As depicted in Figure 1.3, the relationship between research themes will ultimately lead to an iterative process of elicitation of (i) requirements, (ii) technology developments, and (iii) assessment. This loop is typically fueled by input from the utilities and industry partners, both in the requirements and assessment stages. A two-tier approach could then be taken in developing demonstrations and prototypes: (i) Technology and simulation laboratory stage (in various labs), and (ii) Integration tests, prototype deployment, and demonstrations at a real microgrid.

### 1.5.1 Theme 1: Operation, Control, and Protection of Smart Microgrids

Deregulation of the electric utility industry, environmental concerns associated with the central power plants, volatility of energy cost, and rapid technological developments of Distributed Energy Resources (DERs) units, have resulted in a significant proliferation of DER units in the utility power systems, mostly at the distribution voltage levels. The recent trends indicate significant depth of penetration of DER units in power systems will be a reality in the foreseeable future. The term DER encompasses the DG, the Distributed Storage (DS), and any hybrid combination of these. The technical and economic viability of the DER units for power system applications has been the main reason for appearance of the active distribution system and the microgrids concept.

Until recently, the concept of microgrids has mainly been discussed in the context of connecting DER units to an existing utility distribution system, while keeping the

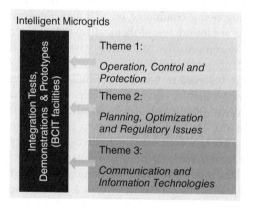

**Figure 1.3** Thematic structure of NSMG-Net research.

operational, control, and protection strategies, and even the infrastructure, of the distribution system intact. Based on this philosophy, a microgrid is defined as a cluster of loads and DER units, serviced by a distribution system, capable of: (i) operation in the grid-connected mode, (ii) limited operational capability in the autonomous or islanded mode, and (iii) transition (under special conditions) between the two modes. In this context, most RD&D activities have been mainly devoted to the control, islanding, and synchronization issues of individual DER units.

However, current trends and the anticipated significant influx of various types of DER units, including controllable loads, has largely outdated the notion of the DER "connection" concept and replaced it with one of "integration." This integration has inspired the concept of an "active distribution network," in which the DER units and controllable loads within the microgrid can experience pre-specified degrees of centralized control and management in response to disturbances and/or the market signals. Thus, the microgrid is viewed as a unified "power cell" that not only requires inter-microgrid protection, control, and management, it also necessitates consider-ations for the intra-microgrid control and performance indices, with respect to the interface hub between the microgrid and the main grid. Since this new concept will have profound impacts, beyond merely electrical boundaries of the microgrid, on the host grid, particularly as the number of microgrids within the host grid increases, regulatory issues and policymaking aspects are also paramount to successful implementation of new microgrid concepts. Furthermore, the developing trends in planning and operation indicate that microgrids should also provide optimal generation in response to market signals and the maximum utilization of renewable generation units.

Transition from the basic microgrid concept, which mainly covers connection and limited control of individual DER units, to the concept of microgrid as a power cell is best achieved based on the use of information and communications technologies (ICT) that enable monitoring, diagnostics, protection, command and control, and power manage-ment. The microgrid cell, including functionalities based on ICT, constitutes the "smart microgrid."

Theme 1 could potentially address the challenges, novel strategies, solutions, and tech-nologies for monitoring, protection, control, and operational strategies, within the regu-latory frameworks and the standards of the smart microgrid. It could primarily focus on the smart microgrid in the application context, and deal with urban, rural, and remote smart microgrid configurations.

The proliferation of DG units in the electric power system (EPS), mainly at the distribution voltage levels, is normally regarded as the main impetus for the formation of a microgrid. The high depth of penetration of DG units not only imposes opera-tional, control and protection challenges, it could also adversely affect the main grid. These effects are caused by: (i) the characteristics of the DG units, e.g. intermittent nature of renewable such as wind and solar photovoltaic (PV) units; (ii) significantly different response times of different DG technologies to control commands and fault scenarios; (iii) low inertia and susceptibility to large variations from the initial operating conditions; (iv) a significantly high degree of unbalanced conditions due to the intrinsic nature of loads, single-phase DG units and the distribution lines; (v) bidirectional power flows, in contradiction to the original design philosophy of the existing distribution systems; (vi) the need to accommodate the plug-and-play operational concept at least for a subset or the smaller size DG units, electric vehicles (EVs) and plug-in hybrid vehicles; and (vii) the requirements to respond to market signals. These challenges manifest themselves differently, and with different degrees of severity and priority, for urban, rural, and remote microgrids.

These technical issues become especially pronounced in those utility grids that operate a large number of weak, long, radial distribution systems or feeders. Many utilities across the globe have little or no experience operating microgrids. The lack of operational experience and inadequate exposure to the new technologies and strategies for operation, control, and protection remain as a major obstacle to increased deployment of DG units, and particularly to utilizing renewable resources in the electricity sector. These obstacles will be further compounded as a result of integration of electric vehicle and plug-in hybrid vehicle units.

Based on pervasive use of ICT, Theme 1 may deal with (i) RD&D of operational strategies for urban, rural, and remote microgrids; (ii) integration strategies of DG units with different operational characteristics, particularly renewables in remote microgrids; (iii) research and development of strategies, and identification of technologies to enable real-time disturbance detection, status monitoring, diagnostic signal generation, and adaptive protection; (iv) development and demonstration of centralized, decentralized, and hybrid control strategies and digital algorithms for stable and reliable operation of the DG units within the microgrid (inter-microgrid) and/or the microgrid as a single entity with respect to main grid-hub (intra-microgrid); (v) provisions for microgrid secondary and tertiary voltage/frequency control; (vi) research and development of power-management strategies for optimal operation of the microgrid; and (vii) system planning and development of operational strategies, including utilization of new technologies, e.g. storage technology and electronic circuit breakers, to resolve issues associated with the very high depth of penetration of DER units, e.g. intermittent renewable, Electric Vehicles (EV) and Plugin Hybrid Electric Vehicles (PHEV) units. Theme 1 could further be sub-divided into the following four research topics.

### 1.5.1.1 Topic 1.1: Control, Operation, and Renewables for Remote Smart Microgrids

This topic may be built around innovative robust control methods, operational strategies, and the corresponding digital algorithms, based on the use of information and communication technologies, with special focus on maximizing the depth of penetration of renewables, and providing reliable operation for remote microgrid configurations. The main research challenges of this topic include:

- Development of a robust and fault-tolerant control strategy that can reliably provide optimum utilization of renewables in remote microgrids which are subject to frequent disturbances and load changes.
- Development of a supervisory control system that could dispatch and maximize depth of penetration of DER units, including renewables in remote microgrids.
- Identification and selection of reliable ICT and the appropriate back-up strategies and algorithms to ensure reliability of supply.

### 1.5.1.2 Topic 1.2: Distributed Control, Hybrid Control, and Power Management for Smart Microgrids

This topic may focus on developing novel ICT-based control and power-management strategies to maintain stability and optimal operational conditions, considering market signals, and to enable operation of a smart microgrid in the grid-connected mode, islanded mode, and transition between these modes. The Virtual Power Plant (VPP) operational mode, under grid-connected conditions of the smart microgrid, could also be a major focus of this topic, which operates in the urban and rural smart microgrid configurations. This topic could also address coordinated control and operation of multiple smart microgrids that shared a single hosted power system. The research focus of this topic includes:

- Generalization of the ICT-based decentralized robust control method to simultaneously accommodate a large number of generation and storage units with significantly different time responses and operational constraints.
- Provisions to guarantee adequate robustness limits of the decentralized controls subject to the smart microgrid intrinsic unbalanced conditions, with a wide range of changes in the size, number, and characteristics of its generating units.
- Development of a power-management system to provide optimal operation of the smart microgrid under various modes of operation with respect to different time frames, for example secondary and tertiary controls in the islanded mode.

### 1.5.1.3 Topic 1.3: Status Monitoring, Disturbance Detection, Diagnostics, and Protection for Smart Microgrids

This topic may look into: (i) novel strategies, algorithms, and technologies, based on use of ICT, to enable real-time disturbance detection, monitoring, and diagnostic signals for urban and rural smart microgrid systems; and (ii) developing novel adaptive protection strategies, algorithms, and the corresponding implementation technologies for the smart microgrid. The main research challenges of this topic could be:

- Development of sensory methods and identification of the corresponding technologies for monitoring the smart microgrid, and the development of computational strategies for real-time extraction of the information for protection.
- Development of an adaptive protection strategy that can reliably identify fault attributes in real-time, subject to a wide variety of the smart microgrid configurations, topological changes, and operational modes.
- Coordination and integration of the smart microgrid adaptive protection strategy with the fast controls of electronically interfaced DER units.

### 1.5.1.4 Topic 1.4: Operational Strategies and Storage Technologies to Address Barriers for Very High Penetration of DG Units in Smart Microgrids

This topic may potentially concern itself with the strategies and the technologies needed to address issues associated with very high depth of penetration of DG units, e.g. Combined Heat and Power (CHP) units and roof-top solar PV units, based on the use of energy storage, ICT, and power electronics in urban, rural, and remote smart microgrids. The main research challenges of this topic include:

- Identification of technical issues, constraints of the existing standards and guidelines, new regulatory issues and policy requirements associated with the very high depth of penetration of DER units in the smart microgrid, and the integration of large number of urban and rural smart microgrids in the power system.
- Research and development of the models and the corresponding analytical and simulation tools for assessment of the various electrical and thermal issues and their

impacts on; (i) operational strategies; (ii) economic considerations; and (iii) regulatory and standard requirements; (iv) communication networks and information technology infrastructure; and (v) the need for advanced technologies; e.g. energy storage systems and electronic-based circuit breakers.

- Research and development of strategies to maximize overall efficiency of the smart microgrid and the host power system, subject to reliability constraints, power quality limits, and environmental requirements and issues.

### 1.5.2 Theme 2 Overview: Smart Microgrid Planning, Optimization, and Regulatory Issues

A microgrid is defined in this theme as a cluster of DERs that are locally controlled and behave, from the perspective of the main electricity grid, as a single electricity producer or load. Some of the key challenges of integrating microgrids into the electricity supply system include the economic justification of building and operating a given microgrid, the operation of multiple microgrids within the main electricity grid, the management of the energy flow between the microgrid and the main grid, and the development of the tools required to facilitate the deployment and operation of microgrids. The topics in this Theme may draw upon past research on components, systems, and operating strategies that are regarded as the building blocks of microgrids.

The issue of the economic justification of integrating DG, including generation based on renewable energy resources, into the distribution grids, has been examined from the point of view of quantifying benefits for and allocating benefits to the different stakeholders, including the owners of the units and the distribution grid operator. Benefits include contribution to the energy supply and energy security, the supply power quality, and environmental impacts. The principles presented in these studies need to be extended to microgrids, and to include ancillary service provisions, such as reactive power generation, voltage regulation, reserve, and black start capability. These services are more readily implemented in microgrids than in individual DGs.

The work on quantifying benefits needs to be extended to multiple microgrids operating within a given distribution network. The interaction of multiple microgrids introduced a new set of operational issues, as it involved a wide range of time frames, from the very fast transient encountered in generation and load switching to the slow transients associated with generation scheduling and the long-term stability of the electric grid. Energy management of loads within the microgrid is essential to ensure energy supply security. Energy management needs also to be considered for the microgrid operating in an autonomous manner or when exchanging power with the main grid. The impact of DG, particularly renewable energy systems (wind, solar) on electricity markets and particularly nodal pricing, was investigated by numerous researchers in the field. Such work needed to be extended to include microgrids, particularly those incorporating renewable energy, to exploit the flexibility they offer.

Furthermore, planning issues of DGs in distribution systems in the context of deregulation have been looked into, as well as the impact of feed-in tariffs and carbon taxes on distribution systems, particularly on DG investments in the long-term. The concept of DG goodness factors was developed to understand the effect on the system of DG power injected at a bus. Researchers also developed a comprehensive hierarchical dynamic optimization model to design and plan distribution systems wherein, in addition to planner-controlled investments, uncoordinated investor DG investments were also included, for which recommendations could be made. Beyond planning, this model's usefulness was also extended into examining the impact of different energy policies on system operation and economics.

There has been extensive work done on modeling DG, particularly systems based on renewable energy, and on large generation stations based on renewable energy, particularly wind farms and more recently solar farms. Such work in turn drew on previous work associated with the modeling of power electronic systems for electricity transmission (High Voltage Direct Current [DC]) and for power system compensation. In addition, a significant amount of work was carried out to develop benchmarks for electricity distribution systems to allow incorporation of DG. There is a need to extend such work to develop the tools to model microgrid components, to benchmark typical microgrid topologies and configurations, and to produce case studies that will facilitate the deployment of microgrids. There is significant interest in validating these tools and approaches using demonstration sites.

This theme could potentially deal with the operation of the microgrid from an overall perspective, and from the point of view of its insertion into the main grid. It should cover issues related to the economic and technical justification of the creation of a microgrid, the interaction with the main grid, including the utility regulatory requirements for connecting to the main grid, and the energy and supply security considerations related to the loads served by the microgrid, including energy management, demand response and metering requirements.

In order to justify the development of a microgrid and facilitate its implementation, metrics should be developed for microgrid assessment and study cases assembled. The network approach is typically needed to pursue research work resulting in the creation and implementation of a microgrid, to develop the tools to assess its economic viability, its benefits for the loads served, and to ensure that its insertion into the main grid could meet the requirements and expectations of the main grid operator.

The theme could, therefore, address issues related to the overall technical and economic justification of the microgrid and its interaction with the main grid. It could examine the general issues of the impact of the microgrid on the main grid, including possible contributions to enhancing the security of the main grid and providing ancillary services, and the requirements of the grid operator regarding the operation of the microgrid to ensure a secure interconnection.

Direct benefits to the stakeholders of this theme would be in providing tools to quantify the benefits of implementing microgrids, not only to the operator of the microgrid, but also to the operator of the distribution grid to which it is connected. Indirect benefits would be the ability to integrate larger amounts of renewable energy, resulting in a reduction of green house gases (GHGs), the lowering of the rate of expansion of the distribution grid and the reduction of losses in distribution grids, due to the existence of local generation, an added flexibility and reliability of the energy supply of the loads served by the microgrid.

Utilities participating in this theme may be convinced to invest in investigating the impact of the deployment of microgrids, particularly on a large scale, on their distribution grid, in particular the positive impacts, energy security, and power quality (system voltage profiles and voltage regulation). They need to know what measures should be taken to integrate this new entity.

### 1.5.2.1 Topic 2.1: Cost–Benefits Framework – Secondary Benefits and Ancillary Services

Increasing the level of complexity of a given system can be justified by the new functionalities or improved performance inherent in the evolved system. In economic theory, this line of argumentation is quantified by attaching monetary values to the cost of the additional complexity, weighed against the monetized benefits that are realized through these enhancements. For utilities and business owners, this analysis is required to develop the business case for a given technology, such that its integration might be justified to stakeholders or a regulatory body. This topic should focus specifically on the cost–benefit framework for microgrids and their feasibility as a new distribution system technology, taking into account the following research challenges:

- Establishing a comprehensive list of all primary and secondary benefits and the framework for quantifying the benefits, including ancillary services;
- Developing a methodology for quantifying and allocating the monetary value of direct and indirect benefits, whatever the stakeholders; and
- Developing a methodology for planning and optimizing the implementation of mechanisms to exploit the operational and financial value of the benefits.

### 1.5.2.2 Topic 2.2: Energy and Supply Security Considerations

This topic may deal with the integration of multiple urban and rural microgrids in the interconnected power system. It should focus on evaluating and quantifying the impacts of the integration of multiple microgrids on the host interconnected power system in terms of (i) technical performance, (ii) reliability of supply, (iii) economical aspects, and (iv) potential infrastructure needs, in particular the data communication supervisory network and the information technology. This topic could potentially provide a host of remedial solutions to address the technical issues, the market requirement, and the standards and regulatory aspects of a power system that incorporates multiple microgrids in the presence of the following research challenges:

- Quantifying the impact of a large penetration of microgrids on the steady state, dynamic, and transient performance of the interconnected power systems;
- Determining the potential violations and/or infringements of the regulatory requirements and standards, and the new guidelines that need to be established for a large penetration of microgrids; and
- Developing supervisory control and power-management strategies and algorithms to enable partial and/or full autonomy of microgrids and to minimize adverse impacts within microgrids

### 1.5.2.3 Topic 2.3: Demand Response Technologies and Strategies – Energy Management and Metering

Given the fact that the concept of a smart microgrid or smart power grid is relatively new at the time of writing this book, this topic should attempt to develop a communication platform for the energy management system (EMS) for microgrids that is reliable, cost-effective, and easily scalable, with the understanding that the intelligent grid requires a reliable and cost-effective communication system that is bi-directional; i.e. to and from the customer for demand response management, allowing various energy sources to be monitored and controlled.

The research network should, therefore, aim to address key aspects of such issues by considering the role of the smart microgrid in the design of modern grids. The basic assumption behind that thinking is that microgrids could act as self-sustaining clusters of generation and customer loads that could interact with the electricity grid and transition from grid-connected mode to an islanded mode; i.e. microgrid that is temporarily disconnected from the main electricity network. In such a case, advanced control strategies and EMSs are fundamental to intelligent microgrid research. The management of building and customer loads and the DERs need to be studied in order to assess the cost and benefits of smart microgrids.

EMSs typically use advanced control and communication technologies to send signals to load clusters that are targeted during a critical or peak demand period. In microgrids, the demand responsive loads, which could be activated upon notification, would be integrated into the optimization strategy. This information would then be conveyed to components responsible for distributed automation of the microgrid.

The transfer of such information would take place through the appropriate communication protocols for the purposes of managing the microgrid during transitional steps between islanded and grid-reconnection modes. Rapid and automatic demand response, coupled with on-site energy generation, need to be evaluated to quantify the range of energy savings from peak shaving and what additional provisions are required to operate the system close to the margins while maintaining similar power service reliability.

The topic should consider optimal asset management through demand response, peak shaving, outage restoration; and the attribution of benefits by assessing market-pricing approaches for microgrid investors as a function of regional or nodal network constraints. In summary, this topic should focus on the following research challenges:

- Determining the environmental, economic, and social impacts of an increased deployment of microgrids, and their overall sustainability;
- Designing energy-aware scheduling algorithms and determining the cost and benefits within microgrids; and
- Proposing algorithms for optimizing the internal microgrid load balancing capability taking into account the impacts and constraints on the reliability and capacity of the main electric grid.

### 1.5.2.4   Topic 2.4: Integration Design Guidelines and Performance Metrics – Study Cases

This topic should focus on constructing models for microgrid components, connecting electric network layers, renewable generation equipment, and simplified models showing essential control interactions between the microgrids, as well as practical study cases, based on existing demonstration installations used in the deployment of new microgrids. In summary, this topic should deal with the following research challenges:

- Defining the control and communication layers and determining modeling requirements, including technical aspects such as communication latency and spread, failure modes, and durations;
- Developing models appropriate for different operating scenarios in different time frames and establishing the modeling details required, and justifying reduced-order models; and

- Producing relevant and universally usable study cases, to facilitate the deployment of microgrids within existing conventional and new intelligent main electricity grids.

### 1.5.3 Theme 3: Smart Microgrid Communication and Information Technologies

Given the fact that nearly 90% of all power outages and disturbances have their roots in the distribution network, the move toward the intelligent grid has to start at the bottom of the chain, i.e. in the distribution system. Moreover, rapid increase in the cost of fossil fuels, coupled with the inability of utility companies to expand their generation capacity in line with the rising demand for electricity accelerated the need to modernize the distribution network through introducing technologies that could help with demand side management and revenue protection. As the next logical step of its evolution, smart grid needs to leverage the existing infrastructure and implement its distributed command and control strategies over the existing structure. The pervasive control and intelligence that embody smart grid have to reside across all geographies, components, and functions of the system. The division between these three elements is of significance as this determines the topology of the smart grid and its constituent components.

A smart grid is defined as a grid that accommodates a wide variety of generation options, e.g. central, distributed, intermittent, and mobile. It empowers the consumer to interact with the EMSs to manage their energy use and reduce their energy costs. A smart grid is also a self-healing system. It predicts looming failures and takes corrective actions to avoid or mitigate system problems. A smart grid uses information technology to continually optimize its capital assets while minimizing operations and maintenance costs. An intelligent microgrid network, which can operate in both grid-tied as well as islanded modes, typically integrates many components. The larger problem in the roll out of a highly distributed and intelligent management system, with enough flexibility and scalability is not only the need to manage the system's growth, but also be open to accommodate ever changing technologies in communication, information technology, and power systems. From a purely information technology perspective, smart grid is often defined as a system in which independent processors in each component, substation, and power plant, work together to deliver system's functionality. These processors are constructed through multitudes of intelligent agents that can work independently, yet collaboratively across many different levels.

This theme should focus on innovative network architectures to support a seamless exchange of data and commands between participants in various transactions across the smart grid network. Provisioning and communication issues between various termination points networked as WANs (Wide Area Networks) for Substation networking, LANs (Local Area Networks) for smart metering and HANs (Home Area Networks) for smart appliances were explored. The topic should address specific research issues such as impact of communication systems on the control system dynamics, reliability, resiliency, and security of network technologies and their associated communication protocols and standards with attention paid to service-level requirements for real-time measurements, data base design, event/alarm processing, and localized intelligence.

The basic assumption made in this theme is that unlike traditional grid-control strategies based on hierarchical models, smart grid relies on a Distributed System of Command and Control. Given the enormity of data produced by arrays of sensors, smart meters, Intelligent Electronic Devices (IED's), substation equipment, field components, etc., it would no longer be practical to rely on a centralized control hierarchy that would need to have access to this myriad of data collected from across the network to make operational and control decisions. The alternative is thus a Distributed Command and Control System. In such a system, the required intelligence is inserted at the appropriate nodes of the system, where relevant data produced locally would be examined, processed and, based on global and/or local control attributes, the required operational and control decisions for that particular node would be taken. The decisions and their consequences would then be communicated to the upper control layers for information and possible update of global and local attributes.

Distributed Command and Control Systems tend to improve system response time (as there would be no need to wait for upper layer control systems to make decisions based on local conditions), reduce the required communication bandwidth (as there would be no need to transport enormous amount of data across hierarchical interfaces to the central control system) and would optimize communication system roll-out and operational cost (as communication technologies could be optimized for individual control layers based on the trade-offs between response time, throughput, reliability, etc).

An area of particular interest would be distributed intelligence through intelligent agents, envisaged for such applications as EMSs, capable of exerting command and control on relevant nodes of the network. In that regard, innovative anatomies should be researched for intelligent agents as autonomous software entities, locked to target environments, and capable of capturing information from other agents or from their surroundings, and make independent decisions on the actions they would need to take.

Having a fully integrated data and control architecture, demand-response strategies and technologies may be developed that could leverage system wide access to sensors, smart meters, and load-control devices. Another application of interest is Information Portals for Utilities and Subscribers. Educating consumers about the impact of their consumption profile on the environment and the electrical infrastructure is considered an important avenue to address the issue of energy conservation. A primary requirement for this is real-time, intuitively constructed data portals and visualization techniques to demonstrate to consumers what their particular consumption would be at any given point in time. It is understood that a similar portal, with many more capabilities and functions, would be required for utility companies. This would provide utility personnel at various levels of decision making with an accurate picture of their assets, system level service, and customer feedback. Issues in data capture, analysis, presentation, and ease-of use are regarded as pertinent research areas in this theme.

Theme 3 should also support the establishment of a cost effective and efficient communication infrastructure between various components of a smart microgrid. This may be achieved through research into optimal architectures and topologies without being dependent on any particular technology, such as Radio Frequency (RF) (Zigbee, Industrial, Scientific, and Medical (ISM) Band RF, WiFi and WiMax), Powerline Career (including narrowband and broadband power line communication [PLC]) and fiber technologies.

Specific outcomes for Theme 3 may include: (i) optimum topologies for HAN, LAN, and WAN networks; (ii) efficient network interfaces and messaging protocols to minimize traffic; (iii) optimum algorithms for service discovery, dynamic routing, and channel access; (iv) parameters for security, authentication, redundancy, and reliability;

(v) dynamic Quality of Service (QoS) requirements of different information types in smart microgrids; (vi) optimum data base design for consumer and utility data; and (vii) visualization techniques for customer and utility information and interaction portals.

### 1.5.3.1   Topic 3.1: Universal Communication Infrastructure

Protocols and security of critical infrastructure are thought to be dependent on a robust, secure, and reliable communications and networking infrastructure, essential for communications between the sensors and the controller for monitoring and fault detection, as well as for message transfer for integrated data management. In summary, this topic should focus on the following research challenges:

- Media-agnostic topology for HAN, LAN, and WAN networks in smart grid and their associated protocols to support asymmetric communication (real-time, event, and polled) among termination points;
- Seamless exchange of data and command through hybrid technologies within each network (Zigbee, Narrowband PLC, Broadband PLC, Narrowband ISM Radios, WiFi, and WiMax); and
- Robust authentication methods associated with various access functionality based on user security level and efficient encryption of command and data with minimal overheads.

### 1.5.3.2   Topic 3.2: Grid Integration Requirements, Standards, Codes, and Regulatory Considerations

This topic needs to deal with issues of grid integration requirements, standards, codes, and regulatory considerations in smart microgrids. In particular, the characteristics of different information types in smart microgrids should be studied. These are needed for establishing their QoS parameters and to classify their dynamic QoS requirements. This may subsequently be used in further studies of the emerging standards and development of efficient transmission, information processing, and interworking techniques and strategies suitable for a robust communications infrastructure, capable of supporting the integration of smart microgrids. In summary, this topic should focus on the following research challenges:

- Optimum communication technologies for integration of smart microgrids as a function of required transactions;
- Standards for end-to-end messaging, command, and control among integrated microgrids; and
- Efficient protocols for supporting distribution automation within and across integrated microgrids.

### 1.5.3.3   Topic 3.3: Distribution Automation Communications: Sensors, Condition Monitoring, and Fault Detection

This topic should deal with the development of advanced sensor networks together with their associated fault detection/monitoring hardware and firmware that are cost-effective and easily integrated with the field components, IEDs, Inverters, and other equipment installed within the microgrid. In summary, this topic should deal with the following research challenges:

- Technology-agnostic topology for the intelligent sensor network;
- Cost-effective technologies for the realization and integration of an intelligent sensor network; and
- An optimal RTOS (Real Time Operating System) to support a dynamically changing sensor network profile.

#### 1.5.3.4 Topic 3.4: Integrated Data Management and Portals

This topic should focus on data extraction and organization of data. A smart microgrid is thought to be data pervasive. As such, this research should deal with systems which either produce massive amount of data, or would need to have access to already digested, processed, and formatted data. The Data Management technology could be used across the entire microgrid system to handle data, command and control information by various microgrid components. The target outcomes for this theme should be innovative Data Base architectures that would dynamically be scalable, configurable, and optimized, as well as Portals and their associated data presentation and visualization techniques, optimized for various stakeholders for various modes of energy transactions on smart grid. In summary, this topic should focus on the following research challenges:

- Anatomies of highly versatile intelligent agents at various command and control layers within and across smart microgrids;
- Dynamically scalable multi-port data base architecture to support local and remote energy management applications; and
- Platform-dependent architecture for User and Utility Portals with their associate presentation and visualization technologies

## 1.6 Microgrid Design Process and Guidelines

The microgrid design process explains what steps are involved; namely design guidelines in terms of best practices. The design process and the guidelines are a result of NSMG-Net and were developed for the Canadian context, however, the guidelines can be applied to microgrids in general.

The overall process for the microgrid design guidelines is shown in Figure 1.4. It primarily consists of two phases, as indicated, RD&D – phase 1, and implementation and validation – phase 2. Each phase has components that have sub-components of their own. While each component is related with other components and sub-components, a sequential process flow can be identified where some iterations may also be necessary to re-design a few components of the microgrid in the RD&D phase. The guidelines do not cover the deployment phase of the microgrid.

The components of the design process are:

1. *Microgrid benchmark compilation.* Defining a base case or benchmark for the microgrid to be compared with is important, therefore, as a starting point the designers should compile the available microgrid benchmarks for the EPS. The benchmarks should represent the business case as closely as possible. The business cases may be that of remote communities, campus microgrids, military bases, etc.
2. *Microgrid elements modeling.* Develop the models used for microgrid elements in these benchmarks, including generators, energy sources, loads, and local energy storage options.

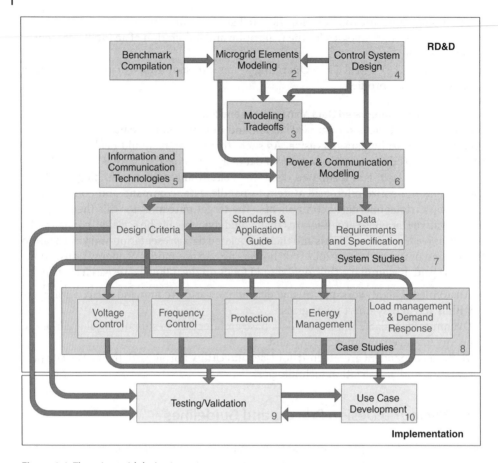

**Figure 1.4** The microgrid design process.

3. *Modeling tradeoffs identification.* Analyze and provide a critical assessment of the modeling detail and the model detail tradeoffs and recommend modeling details as appropriate for the type of study to be performed.
4. *Control systems design.* Design the control systems and approaches implemented in the microgrids used in and meeting the identified requirements.
5. *Information and communication technologies.* Compile the information and communication systems implemented in the microgrids used in and meeting the identified requirements.
6. *Combined power and ICT modeling.* Define a modeling approach that combines the power and communication systems layers, allowing a complete system modeling and study approach.
7. *System studies identification.* Define typical system studies and requirements associated with the operation of the microgrid, in grid-connected and islanded operation, identifying the systems required for case studies, including the appropriate benchmark systems and components.
8. *Case studies development.* Determine a set of system sample case studies required for the study of the various microgrid operating scenarios and contingencies,

for real time operation (voltage and frequency control, protection, islanding and reconnection), and energy management of the generation, storage, and loads within the microgrid, in islanded and grid-connected modes.

9. *Testing and validation.* Test and validate the models with real time Hardware in the Loop (HIL) platforms and field results if possible.
10. *Use case development.* Develop use cases for the sample case studies defined above.

## 1.7 Microgrid Design Objectives

Equally important along with the process flow is to clearly identify the design objectives of the microgrid. A design objective may be a single, or a set of specific targets that the microgrid is needed to achieve. This report lays out six general design objectives that the microgrid may be designed for. The microgrid can also be designed to meet multiple design objectives at a time trying to optimally meet all the design objectives.

The design objectives are:

1. *Environmental footprint reduction.* This design objective is more environment friendly and means operating the microgrid to reduce the greenhouse gas emissions and the carbon footprint of the microgrid.
2. *Network development and increased reliability.* This design objective serves the utility and the customers in terms of a reinforced EPS and a reliable power supply. The utility also benefits by a reduction in the stress on its assets. The design objective can also include a black start option where the microgrid assists the EPS restoration in emergency situations.
3. *Renewable energy integration capacity enhancement.* This design objective allows the integration of variable renewable energy sources without curtailment.
4. *Cost reduction and investment deferral.* This design objective reduces the operational cost of energy by continually optimizing the economical dispatch while serving the consumers.
5. *Efficiency increase.* This design objective reduces the losses of the system by supplying the loads at the point of consumption without going through the transmission which otherwise results in increased losses.
6. *Improvement of operational flexibility and energy dispatchability.* This design objective enables the microgrid to enhance the operational flexibility of the EPS it is connected to. For example, the microgrid could be used to provide ancillary services to the EPS in terms of reactive power support, frequency support, and other primary and secondary response functions.

## 1.8 Book Organization

The rest of the book is organized as follows:

Chapter 2 compiles the available microgrid benchmarks for the EPS, including the models developed for the identified benchmark systems (i.e. a campus type microgrid, a utility type microgrid, and the CIGRE LV North American benchmark). It also identifies and provides justification for the benchmarks used. These benchmarks enable partner companies, end customers, and researchers to work together to develop and qualify various system architectures.

Chapter 3 covers the mathematical modeling of the microgrid components that can be used for microgrid design processes. All necessary components of microgrids were described in details including loads, PVs, Wind, DERs, local DER controllers, and DERs with hardware in the loop.

Chapter 4 analyzes and provides a critical assessment of the modeling detail and the model detail tradeoffs, and recommends modeling details as appropriate for the type of study to be performed.

Chapter 5 compiles the control and protection systems and approaches implemented in the microgrids used in and meeting the identified requirements.

Chapter 6 compiles the information and communication systems implemented in the microgrids used in and meeting the identified requirements.

Chapter 7 defines a modeling approach that combines the power and communication systems, allowing a complete system study.

Chapter 8 defines typical system studies and requirements associated with the operation of the microgrid, in grid-connected and islanded operation, identifying the systems required for case studies, including the appropriate benchmark systems and components. In addition, it covers the overall methodology and design criteria for the Canadian design guidelines.

Chapter 9 determines a set of system sample case studies required for the study of the various microgrid operating scenarios and contingencies, for real time operation (voltage and frequency control, protection, islanding, and reconnection), and energy management of the generation, storage, and loads within the microgrid, in islanded and grid-connected modes.

Chapter 10 summarizes typical use cases for the sample case studies defined in Chapter 9.

Chapter 11 presents the validation of the models with field results of the models presented in prior chapters on the campus microgrid and utility microgrid. Four sample cases from the NSMG-Net project are also presented.

Chapter 12 contains the findings and concluding remarks of the work performed in this report.

# 2

# Microgrid Benchmarks

The first step in the design of a microgrid is to have a representative benchmark model based on the type of microgrid to be designed. Common types of microgrids include commercial/industrial microgrids, community/utility microgrids, campus/institutional microgrids, military microgrids, and remote microgrids [2].

Three different microgrids were used as benchmarks for the analysis of proposed strategies as part of the Canadian national microgrid research network. The benchmark models include the following:

1. A typical campus type microgrid
2. A typical utility type microgrid
3. CIGRE microgrid

Descriptions for the used benchmarks are provided in the following sections.

## 2.1 Campus Microgrid

An intelligent microgrid infrastructure of a typical campus is shown in Figure 2.1 [3]. It enables high-tech companies, end customers, and researchers to work together to develop and qualify various system architectures, configurations, interface protocols, and grid designs to test for and meet national and global priorities for co-generation, efficient transmission and distribution of electricity, load control, demand response, advanced metering, and integration of clean energy sources into the existing and future grids.

This specific campus microgrid is a part of a strategic network [3, 4] that brings together world-class researchers, from many prestigious universities across Canada, to utilize the intelligent microgrid infrastructure to test and validate the results of their research. Together with the network's research, the infrastructure acts as an ideal integration and testing facility, enabling researchers to pool their diverse expertise and resources to develop smart grid technologies and solutions which then need to be customized for the different climate, geographic, and economic demands found across Canada.

### 2.1.1 Campus Microgrid Description

The campus microgrid benchmark is of a typical microgrid that is equipped with its own feeds from the local utility, its own local substations and distribution infrastructure,

*Microgrid Planning and Design: A Concise Guide*, First Edition. Hassan Farhangi and Geza Joos.
© 2019 John Wiley & Sons Ltd. Published 2019 by John Wiley & Sons Ltd.

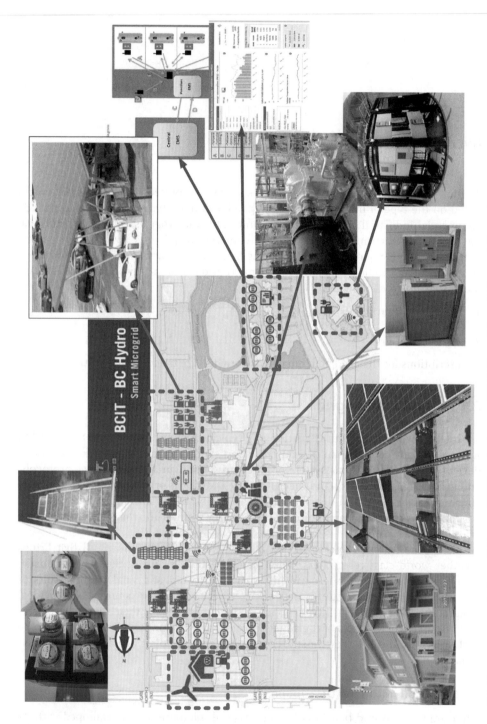

**Figure 2.1** The campus smart microgrid. Source: BCIT Burnaby Campus, NSERC Smart Microgrid Research Network, www.smart-microgrid.ca/.

and its own co-generation capabilities. The utility's substations which feed the campus are equipped with smart components to monitor consumption, demand profile, and distribution yield. Networking of these substations is achieved through their inclusion in the campus's networking engineering lab, which includes communication servers to allow tests on the resiliency of different network topologies, architectures, and protocols for substation networking.

The network engineering lab contains network hardware and test equipment including network routers and switches, traffic generators, impairment emulators, and network analysis tools capable of emulating and testing both small and large-scale network configurations. The architecture of the lab allows for a range of traffic levels and interconnection types so that different grid topologies can be experimented with in a realistic yet controlled environment. The campus intelligent microgrid is an RD&D (Research, Development and Demonstration) platform where existing and future technologies in telecommunication, smart metering, co-generation, and intelligent appliances are employed to develop and qualify the most robust, cost-effective, and scalable solutions required to facilitate and nurture the evolution and emergence of the smart grid in one form or another. The associated university has assembled a consortium of private industry partners, to help with the design and implementation of the intelligent microgrid. The consortium is composed of local and international technology companies at the forefront of technology development in this field and covers a wide array of hardware, software, and system technologies required for the realization of an *"intelligent"* grid. The validation and qualification of architectures, models, and protocols developed are guided, supervised, and scrutinized by partnering utility companies, which are the end-customers of the developed technologies and solutions.

In addition to the termination points, localized co-gen plants (solar, wind, and thermal) are integrated into the microgrid to allow for the development of command and control models for co-generation and integration of such alternative energy sources into the future *"intelligent"* grid. A network of smart terminations and control components is installed on the campus. Buildings are equipped with smart meters, appliances fitted with smart controls, and all these components are fitted with communication modules to enable them to communicate with data aggregation units in substations.

### 2.1.2 Campus Microgrid Subsystems

The following subsections provide details of the components and subsystems present in the campus microgrid.

#### 2.1.2.1 Components and Subsystems

- *The Open Access to Sustainable Intermittent Sources (OASIS) subsystem* consists of 250 kW photovoltaic (PV) arrays, 500 kWh Li-Ion batteries, a 280 kW four-Quadrant grid-tied and island-able inverter, and clusters of level 3 fast DC electric vehicle (EV) charging stations and other campus loads. The details, including the architecture, of the subsystem "OASIS" are shown in Figure 2.2.
- *The Steam Turbine Generator (STG) subsystem* consists of a 250 kW steam turbine along with 75 kWh of Li-Ion batteries and a 25 kW grid-tied inverter. There are clusters of shed-able loads that feed off this turbine. The details, including the architecture, of the "STG" subsystem are shown in Figure 2.3.

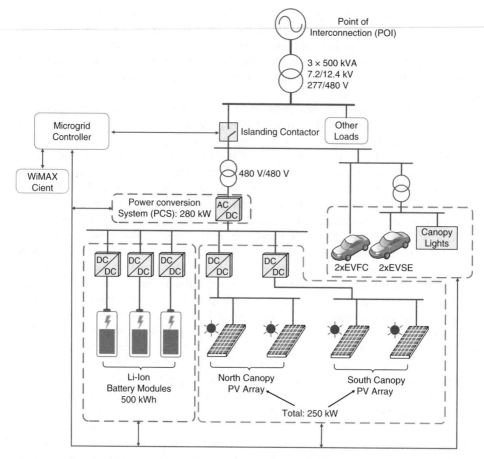

**Figure 2.2** The campus microgrid OASIS subsystem. Source: Project 2.5 Report – Microgrid design guidelines and use cases – Presented at AGM NSMG-Net Sep. 2015.

- *The Gateway Building Microgrid subsystem* consists of 84 solar panels with an installed capacity of 17 kW which feed into the electrical panel on the 2nd floor of the building and includes a power meter that measures the solar generation.
- *The Smart Home subsystem* is a net-zero energy home that consists of 4 kW of PV arrays, 5 kW of wind turbine, 4 kWh of lead-acid batteries, an EV charger, smart appliances, smart meters, in-home displays, smart thermostats, a geo-exchange heating and cooling system, and an energy management system. A power meter measures and records the house's power consumption. The details, including the architecture, of the subsystem are shown in Figure 2.4.

### 2.1.2.2 Automation and Instrumentation

- *Smart Meter Clusters* that measure different buildings across the campus, including its dorms. Most meters are accessible through central power-management software. The clusters make it easier to aggregate the power consumption data and therefore test various load profiles that the microgrid loads offer.

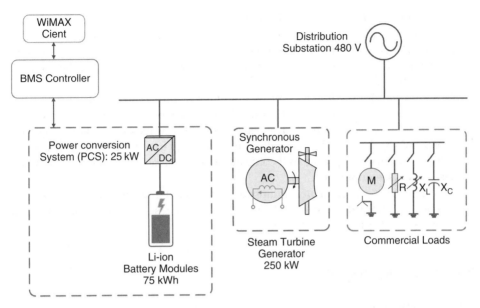

**Figure 2.3** Campus microgrid STG subsystem. Source: Project 2.5 Report – Microgrid design guidelines and use cases – Presented at AGM NSMG-Net Sep. 2015.

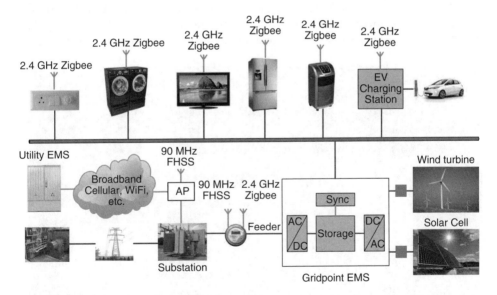

**Figure 2.4** The campus microgrid smart home subsystem smart microgrid. Source: http://www.smart-microgrid.ca/wp-content/uploads/2011/08/Overview-of-the-BCIT-microgrid.pdf.

– *Weather Stations* are installed on campus. These weather stations are all solid state equipment with no moving parts. The weather stations provide accurate local weather measurements, which are invaluable for a microgrid with local renewables to predict the energy generation and energy requirements for weather-dependent electrical loads.

- *The Communication Systems* include Wi-Max 16D 5.8 GHz, as well as a 2.4 GHz ZigBee communication system with base stations and clients spread across the campus. The two systems require some maintenance and upkeep work. Power line communication systems are also provided that allow for networking in the home as well as further distances.
- *The Substation Automation Lab* consists of different IEC61850 compliant relays and intelligent electronic devices (IED's), including a rack of real-time testing hardware as well as signal interface boards. Advanced protection security vulnerability mitigation schemes can be developed and installed on these relays to test the protection logic on multiple vendors' equipment, and confirm the interoperability of the logic between these vendors' equipment.
- *The Microgrid Control Center* equipped with 10 Gbit routers, packet generators, switches, and firewalls provides a laboratory security testing of network connected equipment and systems.

## 2.2   Utility Microgrid

### 2.2.1   Description

The utility microgrid consists of a dedicated 25 kV overhead line that is fed from a 120 kV/25 kV, 28 MVA Y-Δ substation transformer, grounded using a zig-zag transformer. The three-phase system is a four-wire solidly grounded system using 477 AL overhead lines with a length of approximately 300 m from the substation. A

**Figure 2.5** Aerial photograph of the 25 kV Distribution Test Line, showing the physical layout. Source: Ross et al. 2014 [5]. Reproduced with permission of CIGRÉ/Hydro-Québec.

variety of distribution equipment exists on the line that would normally be found in a typical distribution network, including circuit breakers, voltage regulators, shunt capacitor banks, protective switchgear, series reactance, and potential transformers (PTs) and current transformers (CTs) for measurement. The input feeder is connected to three overhead feeders and one underground feeder, which enables system topology

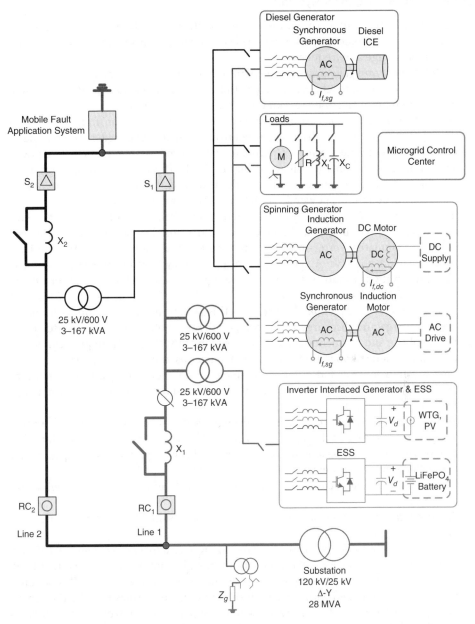

**Figure 2.6** The utility microgrid system. Source: Project 2.5 Report – Microgrid design guidelines and use cases – Presented at AGM NSMG-Net Sep. 2015. Reproduced with permission of Hydro-Québec

reconfiguration to aid in our testing needs. The distributed energy resource (DER), such as generators and loads, can be connected to either Line 1 or Line 2 through three single-phase 14.4 kV/347 V, 167 kVA transformers. The physical layout of the test line is shown in Figure 2.5 and its single-line diagram is shown in Figure 2.6.

### 2.2.2 Utility Microgrid Subsystems

- *400 kVA diesel generator.* The control system is retrofitted to be electrically controlled to enable set-points sent by the microgrid controller to control its active and reactive power, or voltage and frequency.
- *200 kVA induction generator.* This generator is driven by a DC motor whose prime mover is fed from a neighboring feeder. This is used to emulate a wind turbine generator and monitoring of its real and reactive power is required.
- *300 kVA synchronous generator.* This generator is driven by an induction motor with a variable frequency drive, again fed from a neighboring feeder. The microgrid controller can provide power set-points to the governor system and reactive power set-points to the automatic voltage regulator (AVR). The synchronous generator has the ability to operate in isochronous mode, as a synchronous condenser, voltage or frequency droop mode, power set-point operation, and base generator.
- *250 kVA inverter-based generator.* This generator is fed from a neighboring feeder, whose input is rectified to a DC link by a controllable DC power supply. This can be used to emulate inverter-interfaced distributed generations (DGs), such as microturbines, wind, and PVs. A variable power profile may be fed to this DG to emulate such intermittent resources. Output apparent powers are fed back to the controller.
- *100 kWh, 200 kVA energy storage system (100 kWh Li-Ion battery system interfaced with a 250 kVA bi-directional converter).* This system is used to support islanding events, charge and discharge to maximize microgrid objectives, support for other DER, voltage and frequency regulation, assist in transitioning from grid-connected to islanded mode of operation, maintaining power balance, variable real and reactive power, droop control, and other requirements. The microgrid controller should be able to interact with the power conditioning system and the building energy management system (BEMS).
- *300 kW, ±150 kVAR controllable load*, through back-to-back thyristor-based control. Set-points are provided through a PC with a LabView-based graphical user interface (GUI). The loads will be able to follow a set load profile provided by the microgrid controller and provide demand response capability.
- *600 kW load.* Connected to the underground network, and controlled through a similar approach.
- *125 HP induction motor load.* Controlled through the same PC interface and the Labview based GUI.
- *Generic DER controller capable of PQ (real power, reactive power) and /or PV (real power, voltage) control.* Essentially the system should be able to integrate a future DER, assuming its interface is designed to be compatible with the system. This requires the user to be provided with the data format and protocol rules used by the central system in order to interface correctly with the supervisory control and data acquisition (SCADA) system and associated controls. Figure 2.6 shows the microgrid control center of the utility 25 kV distribution test-line.

## 2.3 CIGRE Microgrid

### 2.3.1 CIGRE Microgrid Description

CIGRE has defined benchmark systems for integration of DERs and renewable energy sources (RES) into the North American medium voltage (MV) distribution grid [6, 7]. The benchmark system has an additional DG connected to it to allow it to operate as a microgrid. The system is flexible to allow modeling of both mesh and radial structures. Each feeder includes numerous laterals at which MV/low voltage (LV) transformers would be connected. In North America, radial structures are prevalent, thus single-phase MV lines are included as subnetworks off the three-phase main lines. The nominal voltage on the three-phase sections is 12.47 kV, and on the single-phase sections the line-to-neutral voltage is 7.2 kV. The system frequency is 60 Hz.

The topology of the North American version of the MV benchmark network is shown in Figure 2.7 with Feeders 1 and 2 which operate at 12.47 kV and are fed from the 115 kV

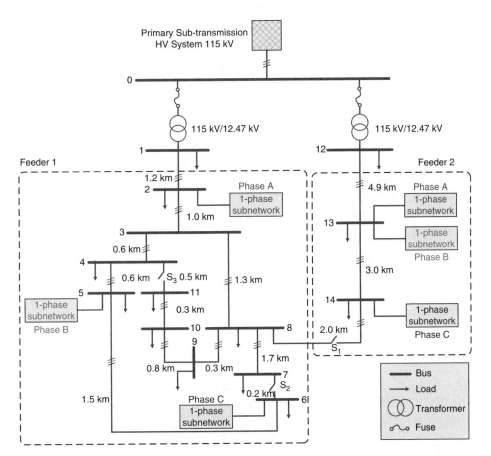

**Figure 2.7** Topology of three-phase sections of North American MV distribution network CIGRE benchmark. Source: CIGRE TF C6.04.02 [7], version 21, August 2010 Reproduced with permission of CIGRE.

sub-transmission system. Either feeder alone or both feeders can be used for studies of DER integration. Further variety can be introduced by means of configuration switches S1, S2, and S3. If these switches are open, then both feeders are radial. Closing S2 and S3 in feeder 1 would create a loop or mesh. With the given location of S1, it can either be assumed that both feeders are fed by the same substation or by different substations. Closing S1 interconnects the two feeders through a distribution line.

Figure 2.8 shows a modified CIGRE benchmark model used by some of NSMG-Net projects. This can be achieved by opening the switch S1 in Figure 2.7. As mentioned the additional DGs are included to allow the network to operate as a microgrid.

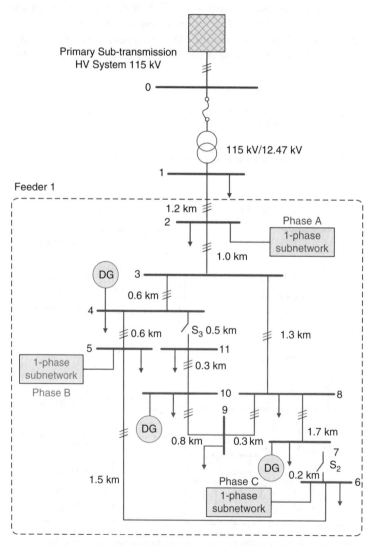

**Figure 2.8** Modified North American MV distribution network CIGRE benchmark. Source: CIGRE TF C6.04.02 [7], version 21, August 2010 Reproduced with permission of CIGRE.

## 2.3.2   CIGRE Microgrid Subsystems

### 2.3.2.1   Load

Table 2.1 contains the values of the coincident peak loads for each node of the benchmark. Note that the load values given for nodes 1 and 12 are much larger than those given for the other nodes. These loads represent additional feeders served by the transformer and are not actually part of the feeder that is modeled in detail.

### 2.3.2.2   Flexibility

Some studies require evaluation of the impact of a DER under varying network conditions. The MV network benchmark offers the following flexibility:

*Voltage*   Testing other voltages (different from 20 kV) on the network is possible by properly readjusting the conductors, conductor spacing, tower configurations, transformers, and other relevant parameters.

*Line Lengths*   The line lengths can be modified as long as voltage drops do not become excessive and a reasonable MV distribution network character is retained.

*Line Types and Parameters*   It is possible to use sections of underground cable or even to use an entire underground network by adjusting appropriately the line parameters instead of using overhead lines. For every change, it would be necessary to modify the line parameters.

**Table 2.1** Coincidental peak loading on the nodes of the CIGRE MV distribution network benchmark.

| Node | R[a] | Apparent power, \|S\| (kVA) | | | | | Power factor (PF) | |
| | | C/I[b] | R | C/I | R | C/I | R | C/I |
| --- | --- | --- | --- | --- | --- | --- | --- | --- |
| 1 | 5010 | 3070 | 4910 | 2570 | 3860 | 3520 | 0.93 | 0.87 |
| 2 | 100 + Subnet | 200 | 50 | 300 | 200 | 300 | 0.95 | 0.85 |
| 3 | — | 80 | 200 | 80 | 50 | 80 | 0.90 | 0.80 |
| 4 | 200 | — | 100 | — | 100 | — | 0.90 | — |
| 5 | 200 | 50 | Subnet | 200 | — | 50 | 0.95 | — |
| 6 | 50 | — | 100 | — | Subnet | — | 0.95 | — |
| 7 | — | 100 | 100 | 100 | — | 100 | 0.95 | 0.95 |
| 8 | 100 | — | 150 | — | — | 200 | 0.90 | 0.90 |
| 9 | 100 | — | 150 | — | 100 | — | 0.95 | — |
| 10 | 150 | — | 100 | — | 250 | — | 0.90 | — |
| 11 | 50 | 150 | 50 | 150 | — | 150 | 0.95 | 0.85 |
| 12 | 1060 | 1260 | 1060 | 1260 | 1060 | 1260 | 0.90 | 0.87 |
| 13 | Subnet | 225 | Subnet | 225 | — | 225 | 0.95 | 0.85 |
| 14 | — | 90 | — | 90 | Subnet | 90 | 0.90 | 0.90 |

a)   Resistive
b)   Capacitive/Inductive
Source: CIGRE TF [7], version 21, August 2010 Reproduced with permission of CIGRE.

**Loads**  Load values can be modified as necessary. If unbalanced loads are desired for the North American MV distribution network benchmark, a load unbalance of $\pm 10\%$ would be reasonable.

## 2.4  Benchmarks Selection Justification

The benchmarks are selected to provide all the components that an active distribution network requires to test and validate various innovative algorithms developed to achieve numerous objectives in the microgrid. The campus and the utility microgrid, both provide an actual physical test bed to form various network topologies that exist in the electric power systems, furthermore the controllability provided over each component provides an opportunity to the designers to study and analyze results from a truly representative system including its technical intricacies. The CIGRE benchmark network was selected to provide the designers a general standard network to test upon. The network model is used by numerous researchers as a test bed for various proposed algorithms and technologies providing the readers and reviewers an independent validation of their proposed techniques. These benchmarks allow studies that include but are not limited to the analysis of the impact of diverse DERs on the power flow, voltage profile, stability, power quality, reliability, as well as the application of methods and techniques of energy management, control, and protection

# 3

# Microgrid Elements and Modeling

This chapter presents the models of the assets that could be used to accomplish the modeling of a microgrid. The models allow the testing of assets (loads, photovoltaics [PVs], wind, distributed energy resources [DERs], DERs controllers, and DERs with hardware-in-the-loop) for unit tests and in a microgrid environment. The inputs required by the models correspond to various functions as required by the advanced control and automation algorithms to be tested. It is also important to independently validate the models in different software simulation, real time simulation, or hardware platforms.

## 3.1 Load Model

Loads impose strong impacts on the distributed energy resource (DER) units and can also dynamically interact with them. Thus, adequate load models are important for studying the dynamics as well as steady-state performance of microgrids. This section presents the structure of two generic load models that can emulate the steady-state and dynamic behaviors of different loads. One is a current source based load model that is capable of modeling the load impedance characteristics. The other model is based on a grid connected inverter that follows the active and reactive power references provided to it. These models are explained in the following subsections.

### 3.1.1 Current Source Based

The structure of the current-based load model is shown in Figure 3.1. The load is modeled by three controlled current sources, one per phase, whose control signals are obtained from a $dq$- to $abc$-frame transformation block [8]. Thus, the $d$- and $q$-axis components of the load current, $i_{Ld}$ and $i_{Lq}$, are dynamically determined based on the load voltage components $v_{Ld}$ and $v_{Lq}$ which, in turn, are calculated from the load terminal voltage $v_{Labc}$. As Figure 3.1 illustrates, a phase-locked loop (PLL) is used to orient the $dq$ frame. Thus, the PLL provides the angle of the load voltage vector, $\theta$. Further, the PLL calculates the angular velocity of the load voltage vector, $\omega_L$, which indeed is the frequency of the load voltage. Due to the function of the PLL, $v_{Lq}$ settles at zero in a steady state.

*Microgrid Planning and Design: A Concise Guide*, First Edition. Hassan Farhangi and Geza Joos.
© 2019 John Wiley & Sons Ltd. Published 2019 by John Wiley & Sons Ltd.

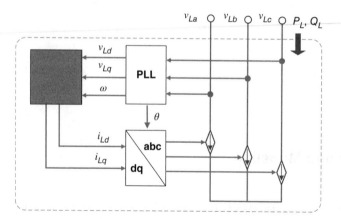

**Figure 3.1** Block diagram of the load model. Source: Haddadi et al. 2013 [8].

The $d$- and $q$-axis components of the load current based on the following dynamic system as shown in (3.1):

$$\frac{d\mathbf{x_L}}{dt} = \mathbf{A_L}\mathbf{x_L} + \mathbf{B_L}\begin{bmatrix} v_{Ld} \\ v_{Lq} \end{bmatrix}$$

$$\begin{bmatrix} i_{Ld} \\ i_{Lq} \end{bmatrix} = \mathbf{C_L}\mathbf{x_L} \tag{3.1}$$

where $\mathbf{A_L}$, $\mathbf{B_L}$, and $\mathbf{C_L}$ are time-invariant matrices that determine the dynamic and steady-state characteristics of the load. Solving Equation (3.1) for $i_{Ld}(t)$ and $i_{Lq}(t)$, one can write:

$$\begin{bmatrix} i_{Ld}(t) \\ i_{Lq}(t) \end{bmatrix} - \mathbf{C_L}e^{\mathbf{A_L}t}\mathbf{x_L}(0) + \int_0^t \mathbf{C_L}e^{\mathbf{A_L}(t-\tau)}\mathbf{B_L}\begin{bmatrix} v_{Ld} \\ v_{Lq} \end{bmatrix} d\tau \tag{3.2}$$

where $\mathbf{x_L}(0)$ denotes the vector of initial states. Equation (3.2) is used to set the dynamic and steady-state characteristics of the load.

### 3.1.2 Grid-Tie Inverter Based

The inverter-based controllable load is designed to absorb the given active and reactive power references. The power references can be obtained from a function dependent on the demanded load output for e.g. the temperature in a heating load. A typical way of modeling a constant impedance base load is given by Equation (3.3).

$$P_{ref}(t) + jQ_{ref}(t) = \frac{V^2_{term}(t)}{Z_{load}} \tag{3.3}$$

where, $P_{ref}$ and $Q_{ref}$ are the active and reactive powers drawn by the load respectively, $V_{term}$ is the terminal voltage of the load and $Z_{load}$ is the constant impedance of the load.

The grid tied inverter that implements the controllable load is shown in Figure 3.2. Decoupled $dq$ control, as shown in Figure 3.3 is used to control the inverter to output the active and reactive power levels.

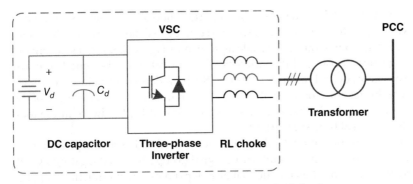

**Figure 3.2** Grid-tie inverter configuration used for controllable loads.

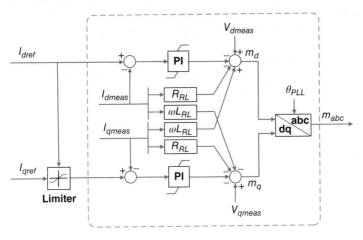

**Figure 3.3** Grid-tie inverter control loop used for controllable loads.

## 3.2 Power Electronic Converter Models

Most of the DERs in microgrids are interfaced to the AC network by a power electronic converter. These converters are conventionally single phase two level, three-phase two level or three-phase three level voltage source converters (VSCs). All these converter topologies consist of the two level half-bridge modules as shown in Figure 3.4. These half bridges are conventionally modeled in three ways according to the detail required for the kind of study conducted [9]. The three types are:

1. *Detailed switch model.* In the detailed switch model based half-bridge, the switches are modeled with their details including the junction voltage drops and the on resistances as shown in Figure 3.5. The switching characteristics of the transistor and the diode is also modeled. Such models provide the most accurate results (includes switching harmonics) but require the most computational resources to simulate. Due to the greater resource requirements such models are used in converter design studies and unit tests where simulating the switching behavior and knowledge of exact harmonics are required.

2. *Switching function model.* In the switching function models, the switches are modeled as a switching state dependent voltage source on the AC side and as a switch state dependent current source on the DC side as shown in Figure 3.6. It does not model the switching characteristics of the switches. Such models provide a good compromise between accuracy and resource economy. Such models are used in system-level studies where the accuracy of the switching behavior and the exact knowledge of the harmonics is not required. These models are more suited for hardware in the loop (HIL) studies with real time simulations.

3. *Average model.* In average models, the switches are modeled as a voltage source on the AC side that depends on the modulating signal (as opposed to the gating signal), and as current sources on the DC side that also depend on the modulating signal as shown in Figure 3.7. Such models are not accurate for frequencies in the switching frequency range, as it does not model switching at all. However, they are the best in performance and are useful for studies of very large networks, or with large time steps where higher order (switching) harmonics are not of interest.

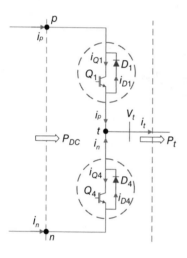

**Figure 3.4** Two level half-bridge topology. Source: Yazdani and Iravani 2010 [9] Reproduced with permission of IEEE/Wiley.

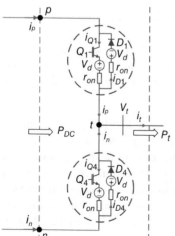

**Figure 3.5** Detailed switch model. Source: Yazdani and Iravani 2010 [9] Reproduced with permission of IEEE/Wiley.

**Figure 3.6** Switching function model.
Source: Yazdani and Iravani 2010 [9]
Reproduced with permission of IEEE/Wiley.

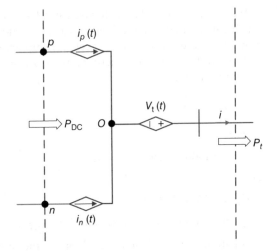

**Figure 3.7** Average switch model. Source:
Yazdani and Iravani 2010 [9] Reproduced
with permission of IEEE/Wiley.

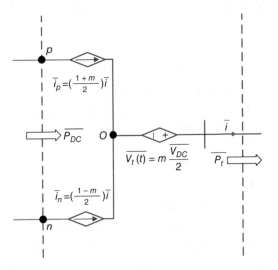

In the figures $V_{DC}$ is the DC link voltage, $Q$ and $D$ represents the transistor and diode respectively, $P_x$ represents the active power, $V_s$ is the AC side voltage, and $m$ is the modulating signal.

## 3.3 PV Model

This section presents a realistic and application-oriented modeling of a Photovoltaic (PV) module. The model is derived on module level as compared to cell level directly from the information provided by the manufacturer's data sheet for PV system simulations. Implementation and results demonstrate simplicity and accuracy, as well as reliability of the model [10]. The model accounts for changes in variations of both irradiance and temperature.

There is a defined relationship between PV module current ($I_m$) and irradiance ($G$) on one hand and between module voltage ($V_m$) and module temperature ($T$) on the other. However, there is no significant relationship between $I_m$ and $T$ or between $V_m$ and $G$.

PV module current, $I_m$ as a function of $G$ at standard test conditions (STC) for a given instant is given as

$$I_{mt} = \frac{G_t}{G_{STC}} I_{STC} = kI_{STC} \tag{3.4}$$

where

$I_{STC}$ = module current at STC;

$G_{STC}$ = 1000 W m$^{-2}$;

$G_t$ = 1 W m$^{-2}$ at time $t$;

$I_{mt}$ = module current at $G_t$; and $k = G_t/G_{STC}$.

Equation (3.4) can also be used for PV module power, $P_m$ as a function of $G$ at STC for a given instant, after replacing $I$ with $P$, as follows:

$$P_{mt} = kP_{STC} \tag{3.5}$$

where

$P_{STC}$ = module power at STC;

$P_{mt}$ = module power at $G_t$.

It is important to note that Equations (3.4) and (3.5) do not account for temperature variations. Therefore, Equation (3.6) is used to account for temperature variations, given as

$$P_{mT} = P_{STC}C_p(T_m - T_r) \tag{3.6}$$

where

$C_p$ = module power temperature coefficient given by data sheet;

$T_m$ = module operating temperature;

$T_r$ = module reference (STC) temperature;

$P_{mT}$ = change in module power at temperature $T$.

Electrical behavior of the PV module with respect to both irradiance and temperature is defined by adding Equations (3.5) and (3.6) together given as Equation (3.7).

$$P_m = P_{mt} + P_{mT} \tag{3.7}$$

where, $P_m$ is the module power as a function of both $G$ and $T$ at a given instant.

In a similar way, the PV module voltage as a function of temperature is given by

$$V_m = V_{OC} + V_{mT} \tag{3.8}$$

where

$V_{mT} = V_{OC}C_v(T_m - T_r)$ = is module voltage at temperature $T$;

$V_{OC}$ = module open circuit voltage at STC;

$C_v$ = module voltage temperature coefficient given by the datasheet.

Finally, dividing Equation (3.7) by Equation (3.8) gives module current ($I_m$) in terms of given $G$ and $T$

$$I_m = \frac{P_m}{V_m} \tag{3.9}$$

The presented PV module model can be easily modeled in any simulation software. Either blocks or functions approach can be used to implement the model. All the values for variables in Equations (3.4) through (3.9) are obtained directly from electrical data and the $IV$ graphs provided in the datasheet.

## 3.4  Wind Turbine Model

The wind turbine is modeled as a controllable current source connected to a grid connected power electronic converter through a Resistive/Inductive (RL) choke and additional filters as shown in Figure 3.8. The wind speed data is required to be input to the maximum power point tracking (MPPT) curves that gives out the maximum extractable power from the turbine. The drivetrain dynamics are modeled by a first order low-pass filter. This power is then fed to the controllable current source to be injected into the grid which is controlled by the rotor side converter. The grid side converter is operated in the DC link voltage control mode. This model allows the mechanical parameters to be modeled in the MPPT curves and in the drive train low pass filter time constant. The model is capable of emulating a wind turbine operating with MPPT or with another power curtailment strategy (as explained in Section 3.7). The wind turbine model explained here is that of a type III wind turbine based on a doubly fed induction generator. The alternate to Type III is the full converter system, Type IV, where the generator is placed behind an AC-DC-AC converter.

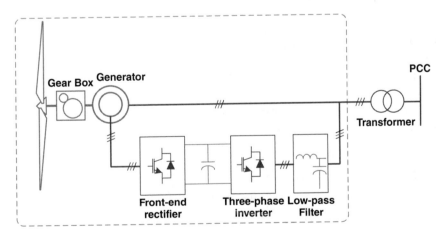

**Figure 3.8** Recommended wind turbine model to be used for microgrids.
Source: Hansen 2012 [11].

## 3.5 Multi-DER Microgrids Modeling

A schematic diagram of a typical radial distribution feeder that is adopted to study a microgrid system is illustrated in Figure 3.9 [12]. The microgrid includes three dispatchable DER units with the voltage rating of 0.6 kV and power ratings of 1.6, 1.2, and 0.8 MVA. It also includes three local loads, and two 13.8-kV distribution line segments. Each DER unit is represented by a 1.5-kV DC voltage source, a VSC, and a series RL filter, and is interfaced to the grid through a 0.6-kV/13.8-kV step-up transformer (with the same power rating as the corresponding DER unit) at its point of coupling (PC) bus. The main utility grid is represented by an AC voltage source behind series R and L elements. The microgrid can be operated in the grid-connected or the islanded mode based on the status of circuit breaker "CB$_g$."

The mathematical model of the microgrid in Figure 3.9 is developed based on its linearized model in a synchronously rotating dq frame. Figure 3.10 shows a one-line diagram of the microgrid. The controller is designed based on the fundamental frequency component of the system. Each DER unit is represented by a three-phase controlled voltage source and a series RL branch. The recommended VSC model for the DERs is either the switching function model or the average model. Each load is modeled by an equivalent parallel RLC network. Each distribution line is represented by lumped series RL elements.

The component models for the test cases are as follows and parameter values are given in Table 3.1.

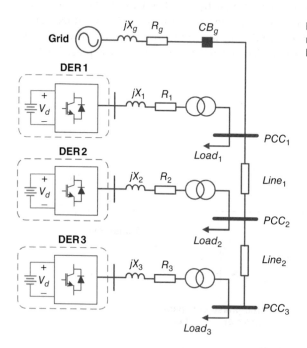

Figure 3.9 Schematic diagram of the multi-DER microgrid system. Source: Etemadi et al. 2012 [12].

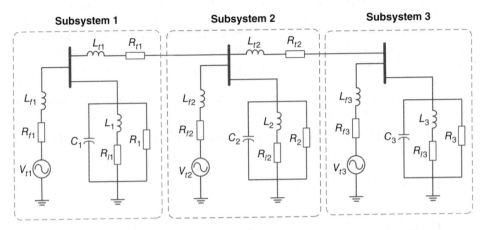

**Figure 3.10** Single-line diagram of the microgrid used to derive state space equations. Source: Etemadi et al. 2012 [12] Reproduced with permission of IEEE.

**Table 3.1** System parameters for the multi-DER microgrid system.

| | | | | | | | |
|---|---|---|---|---|---|---|---|
| *Base values* | | | | | | | |
| $S_{base}$ = 1.6 MVA | | | $V_{base,low}$ = 0.6 kV | | | $V_{base,high}$ = 13.8 kV | |
| *Transformers* | | | | | | | |
| 0.6/13.8 kV | | | $\Delta/Y_g$ | | | $X_T = 8\%$ | |
| *Load parameters* | | | | | | | |
| Load 1 | | | Load 2 | | | Load 3 | |
| $R_1$ | 350 Ω | 2.94 pu | $R_2$ | 375 Ω | 3.15 pu | $R_3$ | 400 Ω | 3.36 pu |
| $X_{L1}$ | 41.8 Ω | 0.35 pu | $X_{L2}$ | 37.7 Ω | 0.32 pu | $X_{L3}$ | 45.2 Ω | 0.38 pu |
| $X_{C1}$ | 44.2 Ω | 0.37 pu | $X_{C2}$ | 40.8 Ω | 0.34 pu | $X_{C3}$ | 48.2 Ω | 0.41 pu |
| $R_{l1}$ | 2 Ω | 0.02 pu | $R_{l2}$ | 2 Ω | 0.02 pu | $R_{l3}$ | 2 Ω | 0.02 pu |
| *Line parameters* | | | | | | | |
| R | 0.34 Ω km$^{-1}$ | | 0.0029 pu | | Line$_1$ | 5 km | |
| X | 0.31 Ω km$^{-1}$ | | 0.0026 pu | | Line$_2$ | 10 km | |
| *Filter parameters (based on DER$_i$ ratings)* | | | | | | | |
| $X_f$ = 15% | | | | Quality factor = 50 | | | |
| *Grid parameters* | | | | | | | |
| $X_g$ | 2.3 Ω | | 0.024 pu | $R_g$ | | 2 Ω | 0.021 pu |

Source: Etemadi et al. 2012 [12]. Reproduced with permission of the IEEE.

- Each DER unit is represented by (i) a constant DC voltage source; (ii) a two-level, insulated-gate bipolar transistor (IGBT)-based VSC; and (iii) a three-phase series filter. The VSC uses a sinusoidal pulse width modulation (SPWM) technique at the frequency of 6 kHz. The VSC filter is modeled as a series branch in each phase.
- The interface transformer for each DER unit is represented as a linear, three-phase, configuration.
- Each three-phase load is represented by a parallel RLC branch at each phase. The quality factor of the inductive branch is 20.

The microgrid of Figure 3.10 is virtually partitioned into three subsystems. The mathematical model of Subsystem 1 in the *abc* frame is as follows:

$$
\begin{cases}
i_{1,abc} = i_{t1,abc} + C_1 \dfrac{dv_{1,abc}}{dt} + i_{L1,abc} + \dfrac{v_{1,abc}}{R_1} \\[2mm]
v_{t1,abc} = L_{f1}\dfrac{di_{1,abc}}{dt} + R_{f1} i_{1,abc} + v_{1,abc} \\[2mm]
v_{1,abc} = L_1 \dfrac{di_{L1,abc}}{dt} + R_{l1} i_{L1,abc} \\[2mm]
v_{1,abc} = L_{t1}\dfrac{di_{t1,abc}}{dt} + R_{t1} i_{t1,abc} + v_{2,abc}
\end{cases}
\tag{3.10}
$$

And the *dq* frame of reference is given as the following:

$$
f_{dq0} = \frac{2}{3}
\begin{pmatrix}
\cos\theta & \cos\left(\theta - \frac{2}{3}\pi\right) & \cos\left(\theta - \frac{4}{3}\pi\right) \\[2mm]
-\sin\theta & -\sin\left(\theta - \frac{2}{3}\pi\right) & -\sin\left(\theta - \frac{4}{3}\pi\right) \\[2mm]
\frac{1}{\sqrt{2}} & \frac{1}{\sqrt{2}} & \frac{1}{\sqrt{2}}
\end{pmatrix}
f_{abc}
\tag{3.11}
$$

where $\theta(t)$ is the phase angle generated by the crystal oscillator internal to DER, or can be derived using a PLL. Based on Equations (3.10) and (3.11), the mathematical model of subsystem 1 in the *dq*-frame is

$$
\begin{cases}
\dfrac{dV_{1,dq}}{dt} = \dfrac{1}{C_1}I_{1,dq} - \dfrac{1}{C_1}I_{t1,dq} - \dfrac{1}{C_1}I_{L1,dq} - \dfrac{1}{R_1 C_1}V_{1,dq} - j\omega V_{1,dq} \\[2mm]
\dfrac{dI_{1,dq}}{dt} = \dfrac{1}{L_{f1}}V_{t1,dq} - \dfrac{R_{f1}}{L_{f1}}I_{1,dq} - \dfrac{1}{L_{f1}}V_{1,dq} - j\omega I_{1,dq} \\[2mm]
\dfrac{dI_{L1,dq}}{dt} = \dfrac{1}{L_1}V_{1,dq} - \dfrac{R_{l1}}{L_1}I_{L1,dq} - j\omega I_{L1,dq} \\[2mm]
\dfrac{dI_{t1,dq}}{dt} = \dfrac{1}{L_{t1}}V_{1,dq} - \dfrac{R_{t1}}{L_{t1}}I_{t1,dq} - \dfrac{1}{L_{t1}}V_{2,dq} - j\omega I_{t1,dq}
\end{cases}
\tag{3.12}
$$

The *dq* frame-based models of subsystem 2 and subsystem 3, are also developed and used to construct the state-space model of the overall system

$$
\dot{x} = Ax + Bu
$$
$$
y = Cx
\tag{3.13}
$$

where

$$x = (V_{1,d}, V_{1,q}, I_{1,d}, I_{1,q}, I_{L1,d}, I_{L1,q}, I_{t1,d}, I_{t1,q}, V_{2,d}, V_{2,q}, I_{2,d}, I_{2,q}, I_{L2,d},$$
$$I_{L2,q}, I_{t2,d}, I_{t2,q}, V_{3,d}, V_{3,q}, I_{3,d}, I_{3,q}, I_{L3,d}, I_{L3,q})^T$$
$$u = (V_{t1,d}, V_{t1,q}, V_{t2,d}, V_{t2,q}, V_{t3,d}, V_{t3,q})^T$$
$$y = (V_{1,d}, V_{1,q}, V_{2,d}, V_{2,q}, V_{3,d}, V_{3,q})^T$$

$$A \subset \mathbf{R}^{22\times22}, B \subset \mathbf{R}^{22\times6}, and\ C \subset \mathbf{R}^{6\times22}$$

Are the state matrices. The system, as defined by Equation (3.13), can alternatively be written as

$$\dot{x} = Ax + B_1 u_1 + B_2 u_2 + B_3 u_3$$
$$y_1 = C_1 x$$
$$y_2 = C_2 x$$
$$y_3 = C_3 x \tag{3.14}$$

where

$$y_i = (V_{d,i}, V_{q,i}), i = 1, 2, 3, \ldots$$
$$u_i = (V_{td,i}, V_{tq,i}), i = 1, 2, 3, \ldots$$

And decentralized controller

$$U_i(s) = C_i(s)E_i(s), i = 1, 2, 3, \ldots \tag{3.15}$$

Equation (3.15) is to be found, where $E_i(s)$ denotes the system error, $U_i(s)$ denotes the input, and the controller transfer function in $C_i(s)$ is restricted to being a proper transfer function $i = 1, 2, 3$. This model (or modeling approach) can be used to model various microgrid structures to perform design and validate centralized or decentralized control strategies [13].

## 3.6 Energy Storage System Model

The model of an energy storage system (ESS) is shown in Figure 3.11. The ESS consists of a grid-tie inverter whose DC-link is fed from a lithium ion battery (or another energy storage). There would conventionally be a DC/DC converter after the battery to ensure a constant DC link voltage and is not shown in the figure for the sake of simplicity. The battery model used was taken from Simulink/SimPowerSystems and the parameters were set in accordance to specified values [14]. The inverter is connected to the grid through an RL choke and transformer. With reference to the current controlled voltage source inverter (VSI) as shown in Figure 3.12, the $i_{dref}$ current reference (corresponding to the active power injection) of the grid tie inverter is dispatched in accordance to the primary control loop set-points. The $i_{qref}$ current reference (corresponding to the reactive power injection) is dispatched in accordance to a primary controller, whose limits are defined by the converter ratings, prioritizing active power. With reference to the voltage

**Energy Storage System**

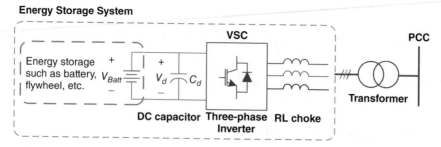

**Figure 3.11** Grid-tie inverter configuration used for the energy storage system.

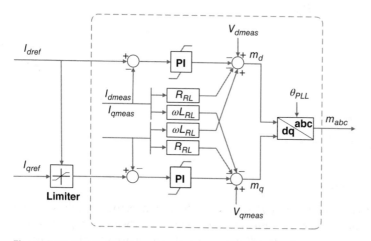

**Figure 3.12** Grid-tie inverter control loops used for the energy storage system operating as a current.

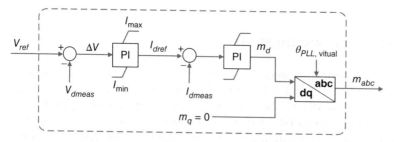

**Figure 3.13** Grid-tie inverter control loops used for the energy storage system operating as a voltage source.

controlled VSI as shown in Figure 3.13, like operation of an uninterruptible power supply (UPS), $i_{dref}$ current reference of the grid tie inverter is dispatched to maintain the reference voltage. A virtual phase-locked-loop is used to set the system frequency. It is important to note that the state-of-charge management control loops have not been incorporated. The microgrid controller will contain this function.

## 3.7  Electronically Coupled DER (EC-DER) Model

The model of an Electronically Coupled Distributed Energy Resource (EC-DER) is shown in Figure 3.14. It consists of a grid-tie inverter whose DC-link is fed from a controllable current source capable of emulating the MPPT curves (Figure 3.15), of the renewable distributed generation (DG), either solar or wind based or, with supplementary functions, emulate curtailment making the source dispatchable. The dispatchability of the DER establishes a reserve for frequency regulation (Delta or Balance control) and for voltage regulation in resistive distribution feeders. The inverter is connected to the grid through an RL choke and transformer. The $i_{dref}$ current reference of the grid tie inverter is dispatched in order to maintain a DC-link voltage. The $i_{qref}$ current reference is dispatched in accordance to a primary controller, whose limits are defined by the converter ratings, prioritizing active power. The overall current controllers of the DERS are shown in Figure 3.16.

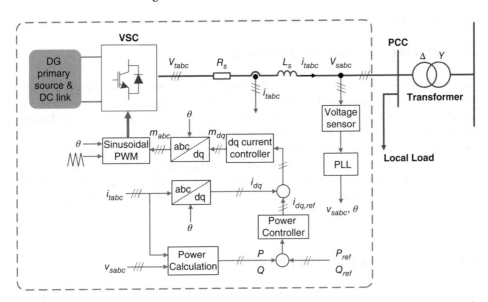

**Figure 3.14** Grid-tie inverter configuration used for inverter-interfaced renewable DGs. Source: Kamh et al. 2012 [15].

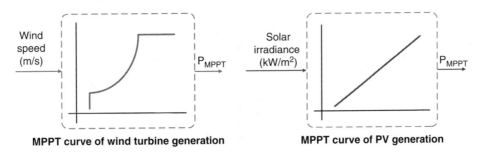

**MPPT curve of wind turbine generation**          **MPPT curve of PV generation**

**Figure 3.15** Look up tables used to emulate the WTG or PV DG.

**DC-link Voltage Controller**     **Grid-tie Controller (Current Controlled VSI)**

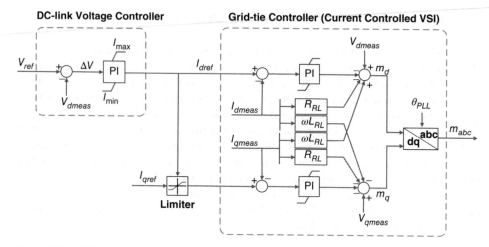

**Figure 3.16** Grid-tie inverter control loop used for renewable DGs.

## 3.8 Synchronous Generator Model

The model of the rotating machine-based generators consists of a synchronous machine fed from a diesel generator. The dynamics of the diesel generator are considered in Figure 3.17. The synchronous machine is connected to the grid through a transformer. The mechanical power $P_m$ is dispatched in accordance to a power reference $P_{ref}$ and speed reference $\omega_{ref}$. The synchronous generator system is also equipped with a DC excitation system that controls the field voltage of the generator $V_f$ in accordance to a reactive power mode and set-points. A combined heat and power (CHP) plant can also be modeled with this model, except that $P_m$ is the mechanical power generated by the CHP plant.

## 3.9 Low Voltage Networks Model

A sequential Monte Carlo (SMC) simulation platform for modeling and simulating low voltage residential networks [16] is presented in this section. This platform targets the

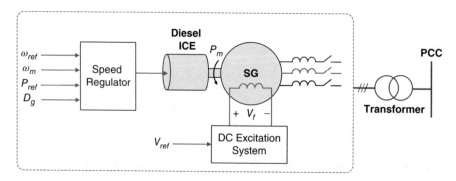

**Figure 3.17** A typical diesel engine synchronous generator system model.

simulation of the quasi-steady-state network condition over an extended period such as 24 hours. It consists of two main components. The first is a multiphase network model with power flow, harmonic, and motor starting study capabilities. The second is a load/generation behavior model that establishes the operating characteristics of various loads and generators based on time-of-use probability curves. These two components are combined together through an SMC simulation scheme.

Since it covers an extended period, a number of unique issues must be addressed, such as when an appliance will be switched-on, how long it will remain in operation, what is the typical solar irradiance level at different snapshots, etc. Therefore, the modeling approach is designed to provide the following features necessary to simulate various distributed management system (DMS) algorithms:

- extended simulation period;
- modeling random behaviors of the loads and generators;
- provide a high resolution output of at least one snapshot per second;
- modeling multiphase networks; and
- modeling of harmonics and motor starting.

To provide the said features the simulation platform models the components as two parts: (i) the electrical model; and (ii) the behavioral model. Once the models are in place, SMC simulations are performed such that the expected value of a function $F(X_{0:t})$ is determined, where $X_{0:t}$ represents the system states from $0{:}t$. The states may be determined using Equation (3.16).

$$E(F) = \sum_{i=o}^{t} F(X_i)p(X_i) \tag{3.16}$$

When used with this modeling approach, the SMC simulation method will determine whether the expected operation status of an appliance at a given instant (or snapshot) is ON or OFF. Such status depends on the previous appliance states (if appliance was ON or OFF at previous instants). Therefore, SMC is applied to create a plausible load profile.

Once such a behavior profile is created, it is combined with the network electrical model to create an instance of the network for that profile. This instance is a deterministic network. Multiphase power flow, harmonic power flow, and motor starting studies are then performed for the instance in a second-by-second sequence. This will yield the network responses over the simulation period for that instance. In reality, there are many instances of load profiles. So the simulation is repeated many times for various randomly generated instances. Statistical analysis is then conducted on the results to obtain statistically valid performance indices.

## 3.10 Distributed Slack Model

This modeling approach involves a distributed slack bus (DSB) model [17], in the sequence-components frame, for power-flow analysis of an islanded microgrid dominated by electronically coupled DER units. The power-flow analysis distributes the system slack among several participating sources since, unlike the grid-connected systems, the reference bus DER unit is anticipated to have a limited power capacity.

The main application of the DSB model is for a fast power-flow analysis of an islanded microgrid to enable its real-time energy management and prevent DER capacity violation in between two consecutive optimal power-flow runs. The DSB formulation is incorporated in a sequence-frame power-flow solver (SFPS).

The formulation has the following features:

- It simultaneously distributes both the real and reactive slack;
- Involves DER units, with different control strategies, to compensate the total slack; and
- Guarantees the power-flow solution to adhere with the DER operating limits.

The approach was found to positively impact the operation of the islanded microgrid by reducing the real-power losses, thus increases the overall system efficiency.

Since the power unbalance (which is defined as the ratio of the negative- plus the zero-sequence power components to the positive-sequence power component) is very small compared to the voltage and current imbalance in the network, it is reasonable to assume that the total system three-phase power slack is approximately three times its positive-sequence counterpart. Hence, the DSB model is defined based on the positive-sequence power-flow, and is incorporated with the positive-sequence power-flow equations of the SFPS, which is where the bulk of the system slack is associated with.

The modeling approach starts with identifying the DER units that will participate in compensating the system slack. Then they are classified as units that can offer to compensate active power slack and those that can offer to compensate reactive power slack. Each unit has a participation factor associated with it, which is defined as the ratio of the slack contribution to the total system slack. The constraints on the participation factors for real and reactive slack compensating units are shown in Equations (3.17) and (3.18).

$$\sum_{i=1}^{m} K_p{}^i + \sum_{i=m+1}^{N} K_p{}^i = 1 \tag{3.17}$$

$$K_q{}^{PV} + \sum_{i=m+1}^{N} K_q{}^i = 1 \tag{3.18}$$

The DSB formulation then models the following power balance equations:
The active power balance equation

$$F_{p1}^i = P_{DER-i}^i + K_p^i P_{slack}^1 + P_{i,comp}^1 - P_{1load}^i - |V_i^1|$$

$$\times \sum_{k=1}^{N} |V_k^1| |y_{BUS-ik}^1| \cos(\theta_i^1 - \theta_k^1 - \theta_{Y_{ill}^1}^1) = 0$$

$$i = 1, 2, \ldots, m, m+1, \ldots, N \tag{3.19}$$

The reactive power balance equation

$$F_{q1}^i = Q_{DER-i}^i + K_q^i Q_{slack}^1 + Q_{i,comp}^1 - Q_{1load}^i - |V_i^1|$$

$$\times \sum_{k=1}^{N} |V_k^1| |y_{BUS-ik}^1| \sin(\theta_i^1 - \theta_k^1 - \theta_{Y_{ill}^1}^1) = 0$$

$$i = m+1, \ldots, N \tag{3.20}$$

The Super PQ bus equation

$$F^i_{Qslack} = 0 = \sum_{i=1}^{N} Q^i_{1load} + \sum_{i=1}^{m} Q^1_{i,comp} + K^{PV}_q Q^1_{slack} - \sum_{i=m+1}^{N} Q^1_{DER-i}$$

$$- \sum_{i=1}^{m} |V^1_i| \sum_{k=1}^{N} |V^1_k||Y^1_{BUS-ik}| \sin(\theta^1_i - \theta^1_k - \theta^1_{Y^1_{ill}}) = 0 \qquad (3.21)$$

The system of Equations (3.19)–(3.21) constitutes the enhanced positive-sequence power-flow equations including the DSB model. This system of equations is solved iteratively using the Newton-Raphson (N-R) method. To obtain a feasible power-flow solution however, the DER operating constraints must not be violated. Thus, real and reactive power limits of the DER units must be imposed subsequent to each SFPS iteration.

## 3.11 VVO/CVR Modeling

This section explains the modeling involved in Volt-VAR optimization (VVO) [18–25] control that optimizes voltage and/or reactive power (VAR) of a distribution network based on predetermined aggregated feeder load profile.

As an important information appliance (IA) in the proposed VVO/conservation voltage reduction (CVR) technique, the VVO/CVR engine is responsible for optimizing voltage, active and reactive power of the feeder accordingly. Thus, the objective of Volt-VAR optimization engine (VVOE) is minimizing the total loss of distribution networks in real time. Equations (3.22)–(3.23) represent the objective function of the VVO/CVR engine. The active and the reactive power loss determination methods based on power-flow equations are given in Equations (3.24)–(3.28).

$$\min\{ S^n_{f,t}\} = \min\{( P^2_{loss,total} + Q^2_{loss,total} )^{1/2}\} \qquad (3.22)$$

$$P_{loss,total} = \sum_{t=1}^{T} \sum_{f=1}^{F} P_{loss,f,t} \qquad (3.23)$$

$$Q_{loss,total} = \sum_{t=1}^{T} \sum_{f=1}^{F} Q_{loss,f,t} \qquad (3.24)$$

$$\sum_{f=1}^{F} P_{loss,f,t} = \sum_{f=1}^{F} \left\{ G_{f,t} \begin{bmatrix} (\alpha_{ij,t} V_{i,t})^2 + (\alpha_{ji,t} V_{j,t})^2 - \\ 2\alpha_{ij,t} V_{i,t} \alpha_{ji,t} V_{j,t} \cos\theta_{ij,t} \end{bmatrix} \\ +G^0_{ij,t}(\alpha_{ij,t} V_{i,t})^2 + G^0_{ji,t}(\alpha_{ji,t} V_{i,t})^2 \right\} \qquad (3.25)$$

where $f = i - j$

$$\sum_{f=1}^{F} Q_{loss,f,t} = \sum_{f=1}^{F} \left\{ -B_{f,t} \begin{bmatrix} (\alpha_{ij,t} V_{i,t})^2 + (\alpha_{ji,t} V_{j,t})^2 - \\ 2\alpha_{ij,t} V_{i,t}\alpha_{ji,t} V_{j,t} \cos\theta_{ij,t} \end{bmatrix} \\ +B^0_{ij,t}(\alpha_{ij,t} V_{i,t})^2 + B^0_{ji,t}(\alpha_{ji,t} V_{i,t})^2 \right\} \qquad (3.26)$$

$$Y_{f,t} = G_{f,t} + jB_{f,t} \qquad (3.27)$$

$$g_{ij,t} + jb_{ij,t} = \begin{cases} \sum_{j \in J} \alpha^2_{ij,t}\ (Y_{ij,t} + Y^0_{ij,t}), i = j \\ -\alpha_{ij,t}\ \alpha_{ji,t}\ Y_{ij,t}, i \neq j \end{cases} \tag{3.28}$$

On the other hand, the required constraints for solving integrated VVO/CVR in real-time intervals are as follows:

1. Bus voltage magnitude constraint

$$V^{min}_{i,t} \leq V_{i,t} \leq V^{max}_{i,t} \xrightarrow{ANSI} 0.95 \text{ p.u.} \leq V_{ir,t} \leq 1.05 \text{ p.u.}$$

2. Active power output constraint

$$P^{min}_{i,t} \leq P_{i,t} \leq P^{max}_{i,t}$$

3. Reactive power output constraint

$$Q^{min}_{i,t} \leq Q_{i,t} \leq Q^{max}_{i,t}$$

4. Active power balance constraint

$$P_{i,t} = P_{Gi,t} - P_{Li,t}$$

$$= \sum_j^J V_{i,t} V_{j,t}(g_{ij,t} \cos \theta_{ij,t} + b_{ij,t} \sin \theta_{ij,t})$$

where $j \in J = (i + \sigma_i)$

5. Reactive power balance constraint

$$Q_{i,t} = Q_{Gi,t} - Q_{Li,t}$$

$$= \sum_j^J V_{i,t} V_{j,t}(g_{ij,t} \sin \theta_{ij,t} - b_{ij,t} \cos \theta_{ij,t})$$

6. DG active power constraint

$$P^{min}_{DGi,t} \leq P_{DGi,t} \leq P^{max}_{DGi,t}$$

7. DG reactive power constraint

$$Q^{min}_{DGi,t} \leq Q_{DGi,t} \leq Q^{max}_{DGi,t}$$

8. System power factor (PF) constraint

$$PF^{min}_{DGi,t} \leq PF_{DGi,t} \leq PF^{max}_{DGi,t}$$

9. Power (thermal) limits of the feeder constraint

$$S_{f,t} \leq S^{max}_{f,t}$$

10. Transformer tap changer constraints

$$\gamma_{tr,t} = 1 + tap_{tr,t} \frac{\Delta V_{tr,t}}{100} = \text{Turn ratio of } OLTC$$

where $tap_{tr,t} \in \{-tap^{max}_{tr,t}, \ldots, -1, 0, 1, \ldots, tap^{max}_{tr,t}\}$ and $\Delta V_{vr,t} = \%\ 3125$

$$\sum_{t=1}^T N_{tr,t} \leq N_{maxtr,t}$$

11. Voltage regulator (VR) constraint

$$\gamma_{vr,t} = 1 + \text{tap}_{vr,t} \frac{\Delta V_{vr,t}}{100} = \text{Turn ratio of VR}$$

where $\text{tap}_{vr,t} \in \{-\text{tap}_{vr,t}^{\max}, \ldots, -1, 0, 1, \ldots, \text{tap}_{vr,t}^{\max}\}$ and $\Delta V_{vr,t} = \% \, 0.625$

$$\sum_{t=1}^{T} N_{vr,t} \leq N_{\text{max}vr,t}$$

12. CB constraint

$$Q_{c,t}^{i} = \beta_{c,t}^{i} \, \Delta q_{c,t}^{i}, \beta_{c,t}^{i} = \{0, 1, 2, \ldots, \beta_{c,t}^{\max}\}$$

13. CB maximum compensation in network

$$\sum_{i=1}^{I} Q_{c,t}^{i} \leq \sum_{i=1}^{I} Q_{L,t}^{i}$$

The aforementioned constraints were added in order to meet the reactive power demand such that the total reactive power provided by the CBs does not exceed the required reactive power of the system. Therefore, the VVOE receives its required real-time data (voltage, current, active and reactive power) from the subscribed IA. From the optimization algorithm, it finds an optimal feasible solution in real time for integrated VVO/CVR (subject to the above-mentioned objective and constraints). Thus, VVOE determines the tap steps of on-load tap changer (OLTC) and VRs. Moreover, it specifies VAR injection points, together with their amounts.

# 4

# Analysis and Studies Using Recommended Models

This chapter presents the modeling details identified as required to perform various kinds of studies on the microgrid. Analysis and assessment of the modeling details along with their trade-offs for the studies are also presented. Table 4.1 presents the required models for all the types of studies typically performed for microgrids. The acronyms used in the table can be found in the list of abbreviations.

## 4.1  Energy Management Studies

Energy management studies are conducted to test various Energy Management System (EMS) operating modes. Since the EMS must communicate with the Electric Power System (EPS) and to manage the microgrid to achieve various objectives while complying with the utility policies and regulations, it requires functional models of all the Distributed Energy Resources (DERs) and loads of the microgrid as shown in Table 4.1. Furthermore, since the EMS operates in all microgrid operation modes (islanded, grid-connected and inter-dispatch operation), all the control loops should also be operational to truly validate the EMS performance. However, phenomena with timescales smaller than the EMS dispatch may not be important for an energy management study.

## 4.2  Voltage Control Studies

Conducting voltage control studies requires testing the microgrid voltage control loops (Volt-VAR control) in both grid-connected and islanded modes of operation. This requires the modeling of controllable loads and the Electronically Coupled Distributed Energy Resources (EC-DERs) along with their Volt-VAR (V-Q) control loops for the grid-connected case when controlling the voltage at the Point of Common Coupling (PCC), and their Volt-Watt (V-P) control loops for the islanded case when controlling

*Microgrid Planning and Design: A Concise Guide*, First Edition. Hassan Farhangi and Geza Joos.

**Table 4.1** Models required for various studies.

| Type of study | DERs | | | | Loads | | Controllers | | | | Power electronic converters |
|---|---|---|---|---|---|---|---|---|---|---|---|
| | | | | | | | | | | | Average (A) – switching function (SF) – detailed (D) |
| | PV | EC | ESS | SG | C | U/C | V-Q | V-P | P-f | P-Q | |
| Energy management | x | x | x | x | x | x | x | x | x | x | A |
| Voltage control | | x | | | x | x | x | x | | | A |
| Frequency control | x | x | x | x | x | x | x | | x | | A |
| Transient stability | | x | | | | x | x | | x | x | A |
| Protection coordination and selectivity | | x | | x | x | x | x | | | | A |
| Economic feasibility | x | x | | | | x | x | | | | A |
| Vehicle-to-grid impact | x | x | | | x | | x | | | | A |
| DER sizing of a microgrid | x | x | x | x | x | x | x | x | x | x | A |
| Ancillary services | x | x | x | x | x | x | x | x | x | x | A |
| Power quality studies | x | x | x | x | x | x | x | x | x | x | SF/D |

the voltage at each DER connection node. The loads required must be modeled with and without controllability to study their realistic impact on the system.

## 4.3 Frequency Control Studies

Frequency control studies are performed to evaluate the performance of the frequency control loops or curves for various primary sources in the microgrid. The scope includes maintaining the power balance of the microgrid in grid-connected and islanded mode using steady-state and transient control including emergency load shedding when required. The control decides the mode of operation of all the DERs and assigns one (or a few) primary sources the responsibility of maintain the network frequency for islanded and for grid-connected modes. For this they require detailed models of all the DERs, loads, and the Volt-VAR (V-Q) and Watt-Hz (P-f) controls.

## 4.4 Transient Stability Studies

Transient stability studies for microgrids look at the phenomenon of the microgrid returning to a stable operating mode after a large disturbance. The focus is on studying the response of DERs to large disturbances in the network including the addition of new DER unit, change in the network topology, change in load dynamics, and increase in DER droop gains.

## 4.5   Protection Coordination and Selectivity Studies

Protection coordination and selectivity studies involve studies to configure the coordination of the protection devices to detect, isolate, and possibly clear a fault from the network as quickly as possible. Therefore, the microgrid protection coordination and selectivity controller must be tested based on the nature of faults to ensure device coordination, correct fault current settings (desensitization of relays), and avoidance of nuisance tripping of relays due to added distributed generation.

## 4.6   Economic Feasibility Studies

Economic feasibility studies are performed on microgrid models to evaluate the economic benefits and for the cost–benefit analysis. These studies may cover technical and economic aspects of microgrids and serve as a prerequisite for modifying and expanding existing network into microgrids as well as building new microgrids. They may also be required to demonstrate improvement in the operation of microgrids when specific microgrid technology, such as an EMS, is implemented.

### 4.6.1   Benefits Identification

Conventionally the objective of an economic analysis of capital expenditure investment is to project the total revenue/benefit realized through capital budgeting techniques to determine the profitability of the investment. However, for microgrid investments, not all cost and benefits may be borne by the investing entity. These benefits may affect stakeholders within the power system such as the utility, microgrid customers, policy makers, and society which may not necessary be the investing actors. Consequently, one needs to carefully identify the cost and benefits of a microgrid implementation in order to justify an investment in a microgrid technology. Once the costs and benefits are identified, they could be correlated with their corresponding stakeholders. The following benefits were considered for the purposes of this project: reduction in energy cost, reliability improvement, investment deferral, minimizing power fluctuation (firm energy), improved efficiency, and emission reduction.

Microgrids may bring a variety of economic, social, and technical benefits. However, quantifying some of these benefits can be complex and challenging: they are either poorly defined or defined differently across jurisdictions. The quantification metrics and methodology used while considering the aforementioned challenges are outlined in the following subsections.

### 4.6.2   Reduced Energy Cost

Microgrid customers and owners could benefit directly from two energy-exchange-related benefits: the locality benefit and the selectivity benefit. Locality benefit is the ability of the microgrid to bypass the upstream network to sell power directly to its loads, allowing it to earn income higher than it would in the wholesale market, and potentially allowing consumers to buy power at a price lower than typical retail costs. The selectivity benefit is a result of the fluctuations in the market and the flexibility of distributed generation (DG) units and controllable loads. The selectivity benefit is not considered in this case since most utilities do not have time- or market-based fluctuation in energy prices. The utility was assumed to have a net metering policy where excess generation to the grid by a microgrid owner is credited against future purchase at a flat rate of 9.9 cents. The net metering policy combined with the utility's tariff rates were used in estimating the cost of energy as outlined in (4.1). The utility charges a flat rate for the first 14 800 kWh of electricity consumed within a month and an additional 5.19 cents for the additional consumption.

The cost of energy $C$ is generally defined by:

$$C = C_{unit} \sum_{t=1}^{y} P(t)dt \tag{4.1}$$

where, $P(t)dt$ is the consumption at time step $t$ and $C_{unit}$ is the unit cost of energy. This is transformed into Equations (4.2) and (4.3) for the purposes of this work, where the coefficients correspond to the utility rate structure.

$$C = 0.016 \sum_{t=1}^{y} P(t)dt + 0.0519u \sum_{t=1}^{y} P(t)dt + 0.0990 \left( \sum_{t=1}^{y} P(t)dt - E_{base} \right) \tag{4.2}$$

$$\left\{ \begin{array}{l} u = 0, \text{if } \sum_{t=1}^{y} P(t)dt \leq 14800 \\ u = 1, \text{if } \sum_{t=1}^{y} P(t)dt > 14800 \end{array} \right\} \tag{4.3}$$

### 4.6.3 Reliability Improvement

Microgrids can improve reliability of the local network due to their ability to operate in an islanded mode when disconnected from the main grid. In the event of islanding, service within the local microgrid is sustained by available local resources in the microgrid. This may include dispatch of additional DER units or energy storage system (ESS), and curtailing or shedding of less critical loads to ensure there is sufficient power to supply more critical loads. Reliability is typically measured by the System Average Interruption Frequency Index (SAIFI) and System Average Interruption Duration Index (SAIDI) indices: SAIFI measures the average number of interruption experienced by customers while SAIDI measures the average duration; they are defined by Equations (4.4) and (4.5):

$$SAIDI = \frac{\Sigma_i N_i}{N_T} = \frac{\Sigma_k \lambda_k N_k}{N_T} \tag{4.4}$$

$$SAIFI = \frac{\Sigma_i r_i N_i}{N_T} = \frac{\Sigma_k U_k N_k}{N_T} \tag{4.5}$$

where $N_i$ is the number of customers affected by interruption $i$, $N_T$ is the total number of customers, $N_k$ is the number of customers at load point $k$, and $r_i$ is the duration of interruption $i$. Considering the microgrid as a single load point, SAIFI and SAIDI simply becomes $\lambda_{\mu G}$ and $U_{\mu G}$. In addition to these indices, the expected Non-Delivered Energy (NDE) can simply be found as:

$$NDE = U_k \times P_k \tag{4.6}$$

where $P_k$ is the average or expected demand at the load point. If there is a possibility that the DER units may have inadequate capacity to meet the load at all times (due to load or resource variation or due to planned or unplanned equipment downtime), an additional term must be considered, which is the Probability of Adequacy (PoA). Nevertheless, in this work, the PoA is dependent on the desired state of charge (SoC) of the ESS. Hence the further away the SoC at time $t$ is from the desired SoC, the less likely it is to sustain service in the event of islanding.

### 4.6.4  Investment Deferral

The distributed network operator or utilities are the main beneficiaries from deferred investment and upgrade costs. Infrastructure investment deferral is highly correlated with reducing peak loading as growth of peak loads can necessitate infrastructure upgrades, along with aging of equipment. An upgrade in equipment which differed to sometime in the future could result in savings to the network operator. Herein, the investment deferral can be determined by comparing the Net Present Value (NPV) or annualized value of a future investment $C_{inv}$ to that of the deferred future as illustrated in Equations (4.7) and (4.8).

$$D = NPV_{base} - NPV_{\mu G} \tag{4.7}$$

$$NPV = \frac{C_{inv}}{(1 + r)^y} \tag{4.8}$$

If the customers also pay for monthly peak power consumption a reduction in their peak usage will result in a reduction of this cost in addition to the deferred investment experienced by the distribution network operator (DNO).

### 4.6.5  Power Fluctuation

Implementation of an EMS can aid smoothing of fluctuations in power generation caused by intermittent resources and contribute to the demand-side management plan of the power grid. The utility is assumed to purchase energy at an agreed upon price via an electricity purchase agreement. The average agreed price for generators with renewable energy resources are higher. The distributed generator will receive a penalty should it fail to maintain output within the pre-agreed range. The opportunity cost, $F$, of providing firm electricity is used in valuing the cost of fluctuation as shown in Equation (4.9):

$$F(t) = (P(t) - P_{avg}(t)) \, C_{pen} \tag{4.9}$$

where $P_{avg}$ is assumed to be the firm energy output.

### 4.6.6  Improved Efficiency

Local generation can also result in improvement in voltage profiles and losses reduction. Electricity generated locally to loads in microgrids should reduce the real power losses in a power network unless the system is poorly designed and/or overloaded. "Losses outside the microgrid" is energy purchased or produced by the utility that will not be sold. "Losses within the microgrid" is energy produced or purchased by the microgrid owner that will not be utilized by the customers/microgrid load or exported upstream. In both cases, the cost of upstream losses is valued at marginal wholesale price and applied as cost to the system operator, and the cost of internal microgrid losses is applied to the Independent Power Producer (IPP) or microgrid owner, being valued at the cost of energy. The microgrid customer is assumed to be the owner as well so the internal losses are applied to the owner in this formulation.

### 4.6.7 Reduced Emission

Microgrids can reduce emissions of certain pollutants based on renewable or low-emission (e.g. natural gas-based) DERs. However, the generation resources of our base case are primarily hydro which makes it intangible to estimate reduction in emission by the microgrid under consideration. Nevertheless, the microgrids' loads are primary electric vehicles, so for every kWh of energy provided to these cars, the society avoids emission from equivalent volume of gasoline consumed by a gas engine vehicle. A survey by Natural Resources Canada (NRCAN) of 2016 electrical vehicle models, showed that the average electric vehicle travels the same distance per kWh as the average gasoline car will do on a 0.404 l of gasoline. Hence the emission cost is determined as outlined in Equation (4.10), where $C_{CO2tax}$ is the $CO_2$ tax per liter of gasoline.

$$C_e = 0.404\, C_{CO_2 tax} \sum_{t=0}^{y} P(t)dt \qquad (4.10)$$

## 4.7 Vehicle-to-Grid (V2G) Impact Studies

Vehicle to grid (V2G) impact studies are also important as the services provided by the vehicles that are not defined well as are the other DERs. The randomness attached to their availability and the non-conventional way of service provision (especially ancillary services) by them warrant detailed V2G impact studies to ensure technical and economic feasibility of the microgrid in both grid-connected and islanded mode of operation.

## 4.8 DER Sizing of Microgrids

DER sizing for microgrids is important to achieve the objectives of the microgrid. The requirement (or the lack thereof) of an ESS and its scheduling in meeting those objectives while meeting the utility standards and requirements is ascertained.

## 4.9 Ancillary Services Studies

The economic and technical feasibility in the provision of ancillary services is ascertained in such studies. Ancillary services may be active or reactive power related, each could be provided separately by separate DERs depending on their capacity and sizes or by the microgrid as a collective contributor to the grid at the point of interconnection.

## 4.10 Power Quality Studies

The power quality studies in microgrids are mainly related with voltage flicker, voltage sag/rise, and harmonics. These phenomena mainly result from the nonlinear characteristics of load, abrupt change of load/generation (e.g. large motor starting, capacitor switching), volatile nature of renewable energy resources, and switching characteristics of power electronic converters. Solutions to these problems include harmonic filters, Static VAR Compensator (SVC), and ESS.

## 4.11 Simulation Studies and Tools

Microgrid studies require various tools to conduct the case studies required in the different phases of development of the microgrid. The time-scale of interest varies significantly with the type of study conducted. For example, an investment planning or an economic feasibility study will require simulations that are able to provide a near accurate picture of the investment in years, whereas the grid code compliance studies would require simulations that are capable of simulating accurately up to tens of microseconds to correctly simulate the harmonics. Various tools are available that allow the user to model and simulate the power system as required for the study to be conducted [26–29]. Table 4.2 summarizes the requirements and characteristics of the modeling platforms for various types of studies performed. Some of the popular tools available to conduct such studies are also presented in the table.

**Table 4.2** Types of simulation tools required to perform studies.

| Types of studies | Studied phenomena Financial (F), Energy (En), Emissions (Em), Technical (T) | Types of simulations used | Timescales of interest |
|---|---|---|---|
| Investment planning | F | | Years |
| DER sizing of a microgrid | En | Monte Carlo with power flow | Years, months |
| Economic operation | En, F, and Em | | Years, months, days |
| Energy management | En and Em | | Days, hours |
| Voltage control | T | Dynamic power system analysis positive sequence phasor type simulations | Minutes, seconds |
| Frequency control | T | | |
| Transient stability | T | | |
| Protection studies and testing | T | Offline and real time electromagnetic transient simulation | |
| Power quality | T | | |
| Unit design test | T | | Seconds, sub-seconds |
| Pre-commission and compliance test | T | Real time electromagnetic transient simulation | |

# 5

# Control, Monitoring, and Protection Strategies

Microgrid control functions operate at various timescales [30]. Grid interactive control functions operate at a much larger timescale than the device-level controls. The relation between the different control functions are shown in Figure 5.1. This chapter covers several control monitoring and protection schemes and strategies that can be used in microgrids to meet the requirements and objectives for which the microgrid is designed. The strategies are classified into Levels 1–3 as defined by the standard IEEE 2030.7 [31]. The classification is shown in Figure 5.2.

## 5.1   Enhanced Control Strategy – Level 1 Function

An enhanced control strategy for electronically coupled distributed energy resources (EC-DERs) that improves the performance of the host microgrid under network faults and transient disturbances [32] is described in this section. The control strategy does not require controller mode switching and enables the EC-DERs to ride through network faults. The enhanced control strategy has the following features:

- It improves the performance of the host microgrid under network faults and transient disturbances.
- When implemented, the host microgrid can ride through network faults, irrespective of whether they take place within the microgrid jurisdiction or strike the upstream grid, and quickly reclaim its pre-fault operating condition.
- The control strategy enables the microgrid to retain its power quality for the duration of the faults, in both modes of operation, which is a desirable property for the detection of certain classes of faults, as well as for sensitive loads.

Figure 5.3 illustrates a schematic diagram of the three-phase EC-DER. The EC-DERs consist of (i) a DC voltage source, which represents a conditioned prime energy source augmented with an energy storage device, connected in parallel with the voltage-sourced converter (VSC) DC-side terminals and DC-link capacitor; (ii) a current-controlled VSC; (iii) a three-phase low-pass LC filter; and (iv) the interface switch $SW$ ensures that the EC-DER unit can be connected to the rest of the microgrid only if its terminal voltage is in phase with the network voltage (this process is referred to as the "local Demand Response (DR) synchronization." The circuit components $L_s$ and $C_s$, respectively, denote the inductance and capacitance of the LC filter, and $R_s$

*Microgrid Planning and Design: A Concise Guide*, First Edition. Hassan Farhangi and Geza Joos.
© 2019 John Wiley & Sons Ltd. Published 2019 by John Wiley & Sons Ltd.

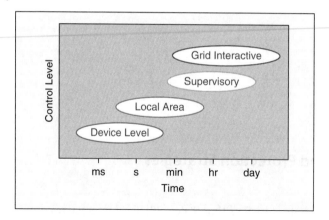

**Figure 5.1** Microgrid control function timescales. Source: Joos et al. 2017 [30].

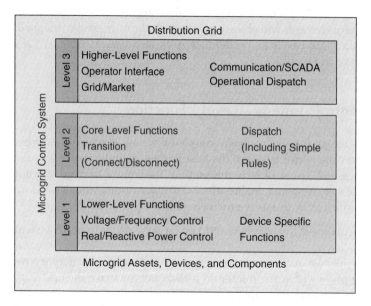

**Figure 5.2** The microgrid control system functional framework core functions. Source: Joos et al. 2017 [30].

represents the ohmic loss of $L_s$ and also embeds the effect of the on-state resistance of the VSC switches. The EC-DER exchanges with the rest of the microgrid (including the interconnection transformer Tr) the real- and reactive-power components $P$ and $Q$. The main control blocks of the EC-DER model are described in the following sub-sections.

### 5.1.1 Current-Control Scheme

The function of the current-control scheme is to regulate the VSC AC-side current, $i_{tabc}$, by means of the pulse width modulation (PWM) switching strategy. Dynamics of

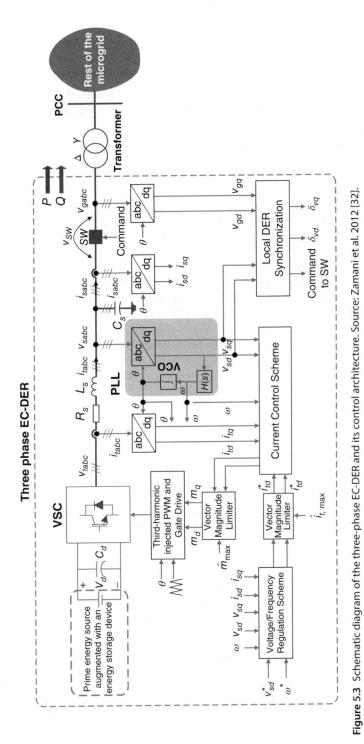

**Figure 5.3** Schematic diagram of the three-phase EC-DER and its control architecture. Source: Zamani et al. 2012 [32].

the *dq*-frame components are governed by the following:

$$L_s \frac{di_{td}}{dt} = -R_s i_{td} + L_s \omega \, i_{tq} + m_d \left( \frac{v_{dc}}{2} \right) - v_{sd},$$

$$L_s \frac{di_{tq}}{dt} = -R_s i_{tq} + L_s \omega \, i_{td} + m_q \left( \frac{v_{dc}}{2} \right) - v_{sq}, \tag{5.1}$$

where $m_d$ and $m_q$, respectively, denote the $d$-and $q$-axis components of the three-phase PWM modulating signal $m_{abc}(t)$. The variable $\omega$ is related to the angle $\rho$ as $\omega = d\rho/dt$.

### 5.1.2 Voltage Regulation Scheme

The objective of the voltage magnitude regulation scheme is to regulate $v_s$, that is, the magnitude of $v_{sabc}$. The regulation of $v_s$, effectively requires the regulation of $v_{sd}$ in the *dq*-reference frame since $v_{sq} = 0$. In a single-unit microgrid, this can be assigned a value equivalent to the nominal magnitude of the network voltage. In a multi-unit system, however, the reference for $v_{sd}$, $v_{sd}^*$ is commonly obtained from the droop characteristic:

$$v_{sd}^* = D_Q (Q^* - Q) + V_0$$

where, $Q^*$ is the setpoint for the reactive power output of the EC-DER in the grid-connected mode of operation and $V_0$ is the nominal network voltage magnitude; $Q$ is the reactive power output of the EC-DER, and the constant $D_Q$ is the reactive droop coefficient.

Dynamics of $v_{sd}$ and $v_{sq}$ are governed by the following:

$$C_s \frac{dv_{sd}}{dt} = C_s \omega \, v_{sq} + i_{td} - i_{sd},$$

$$C_s \frac{dv_{sq}}{dt} = -C_s \omega \, v_{sd} + i_{tq} - i_{sq} \tag{5.2}$$

### 5.1.3 Frequency Regulation Scheme

The objective of the frequency regulation scheme is to regulate $\omega$, that is, the frequency of $v_{sabc}$, at the set-point $\omega^*$. In a single-unit microgrid, $\omega^*$ can be assigned a constant value corresponding to the network nominal frequency, for example, $377 \text{ rad s}^{-1}$ for a 60-Hz power system. However, in a multi-unit microgrid, $\omega^*$ is determined by the droop characteristic

$$\omega^* = D_P (P^* - P) + \omega_0$$

where, $P^*$ denotes the set-point for the real power output of the EC-DER in the grid-connected mode of operation, and signifies the nominal power system frequency. $P$ is the real power output of the EC-DER, and the constant $D_P$ is the droop coefficient.

### 5.1.4 Enhanced Control Strategy Under Network Faults

Figure 5.4 shows the enhanced voltage regulation scheme and Figure 5.5 shows the frequency regulation with phase angle restoration to ensure proper operation of the EC-DER under network faults and severe voltage imbalances. The overall control system with outer voltage control and inner current control loops are shown in Figure 5.6.

**Figure 5.4** Enhanced voltage magnitude regulation scheme for the three-phase EC-DER. Source: Zamani et al. 2012 [32].

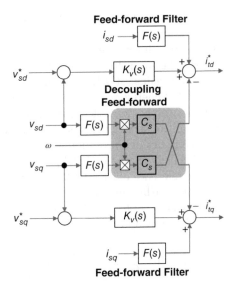

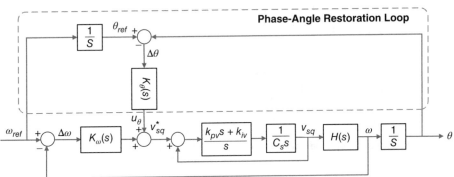

**Figure 5.5** Frequency regulation loop, augmented with the phase angle restoration loop. Source: Zamani et al. 2012 [32].

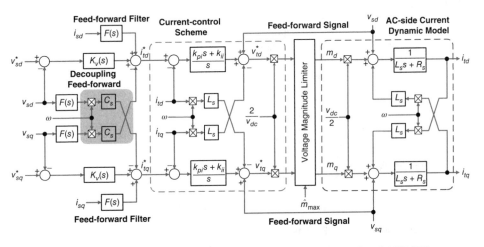

**Figure 5.6** Enhanced voltage magnitude regulation scheme. Source: Zamani et al. 2012 [32].

## 5.2 Decoupled Control Strategy – Level 1 Function

This section introduces a control strategy that enhances the transient performance and stability of a droop-controlled microgrid [33]. The dependency of dynamics on the droop gains, steady-state power flow, and network/load in a droop-controlled distributed energy resource (DER) unit results in poor transient performance or even instability of the network in the event of a system disturbance. A gain-scheduled decoupling control strategy can be used to decouple these dependencies by reshaping the characteristics of conventional droop using supplementary control signals; these control signals are based on local power measurements and supplement the *d*- and *q*-axis voltage reference of each DER unit. The impact of the control on the DER and network dynamics can be studied by calculating the eigenvalues of the microgrid with the control.

The overall control strategy is shown in Figure 5.7 and the decoupled dependencies control is shown in Figure 5.8. The supplementary control is outlined by the following equations:

$$\begin{bmatrix} \delta v_{td} \\ \delta v_{tq} \end{bmatrix} = \mathbf{G} \begin{bmatrix} \Delta i_{td} \\ \Delta i_{tq} \end{bmatrix} + \begin{bmatrix} n\Delta Q \\ mD_p\Delta P \end{bmatrix}$$

$$\mathbf{G} = \frac{3}{2} \frac{V_t^2}{P^2 + Q^2} \begin{bmatrix} -Q & P \\ -P & -Q \end{bmatrix} \tag{5.3}$$

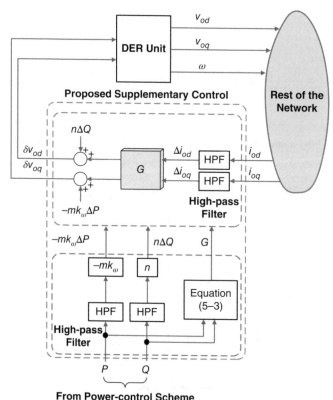

Figure 5.7 The decoupled control strategy with the host DER unit. Source: Haddadi et al. [33].

**Figure 5.8** Block diagram of the dependencies decoupling control strategy. Source: Haddadi et al. [33].

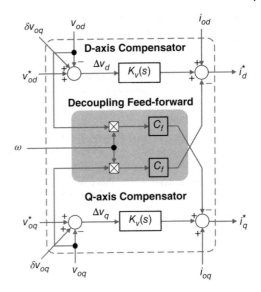

where, $P$ and $Q$ are provided by the power control scheme of the system, $m$ and $n$ are the real and reactive droop gains of the DER unit respectively, the $\Delta$ symbol denotes the small signal perturbation in the output quantities, and $D_p$ is a gain parameter used for the frequency regulation.

## 5.3 Electronically Coupled Distributed Generation Control Loops – Level 1 Function

### 5.3.1 Voltage Regulation

In the case of distribution grids with low X/R ratios, both reactive and active power can be used to regulate the voltage. Using reactive power, volt-VAR control can be employed with (i) voltage control – whereby the Distributed Generation (DG) operates in isochronous mode and sets the system voltage (ii) power factor control – whereby the DG operates at a fixed power factor, and (iii) fixed $Q$ with droop voltage – whereby the DG is dispatched a reference $Q_{ref}$ and droop gain $D_Q$ to regulate the voltage. Voltage rise mitigation is also possible with active power curtailment using a volt-watt function. Using the volt-watt function, a voltage droop is employed to regulate the active power injection when the voltage level reaches it limit, with a droop gain of $D_{Q1}$. The description for each control loop for voltage regulation is summarized in Table 5.1.

### 5.3.2 Frequency Regulation

Additional primary control loops can be incorporated to regulate the system frequency. These control loops typically have a power reference, either (i) delta control – reserves a fixed margin of the available power, (ii) balance control – caps the maximum output power to a specified limit, or (iii) maximum power point tracking (MPPT) control – injects the full power available. A frequency drop gain is then superimposed onto

**Table 5.1** Voltage regulation control loops descriptions for electronically coupled DGs as shown in Figures 5.9 and 5.10.

| Power control loop | Description | Parameters |
|---|---|---|
| Reactive power (Volt-VAR mode) | Voltage control – isochronous mode | $V_{ref}$ |
| | Fixed power factor function | $PF$ |
| | Fixed Q and voltage droop | $Q_{ref}$, $D_Q$ |
| Active power (Volt-Watt mode) | Voltage droop (no contribution for under voltage scenario) | $D_{P1}$ |

**Table 5.2** Frequency regulation control loops descriptions for electronically coupled DGs as shown in Figure 5.10.

| Power control loop | Description | Parameters |
|---|---|---|
| Active power (Freq-Watt mode) | Delta control with frequency droop | %, $D_{P2}$ |
| | Balance control with frequency droop | pu, $D_{P3}$ |
| | MPPT with frequency droop (no contribution for under frequency scenario) | $D_{P4}$ |

the power reference to regulate the frequency. It should be noted that MPPT control will not have a frequency regulation contribution for the under-frequency scenario. The description for each control loop for frequency regulation is summarized in Table 5.2.

The control loops for the reactive and active power for an electronically coupled DG are shown in Figures 5.9 and 5.10 respectively. The selector is used to select the operating mode of the DG. The input and output labels for all the control loops used are listed in Table 5.3.

## 5.4 Energy Storage System Control Loops – Level 1 Function

### 5.4.1 Voltage Regulation

Similar to the inverter interfaced DGs, when operating in grid-connected mode or islanded mode (with rotating generation) the voltage can be regulated with reactive power functions. Using reactive power, Volt-VAR control can be employed with (i) voltage control – whereby the DG operates in isochronous mode and set the system voltage, (ii) power factor control – whereby the DG operates at a fixed power factor, and (iii) fixed Q with droop voltage – whereby the DG is dispatched a reference $Q_{ref0}$ and droop gain to regulate the voltage. The description for each control loop for voltage regulation is summarized in Table 5.4.

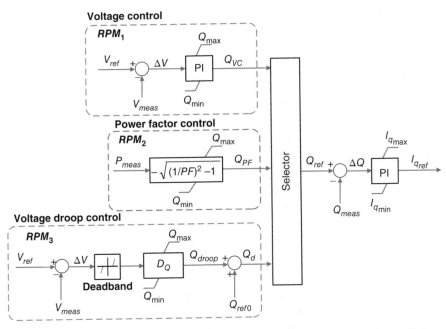

**Figure 5.9** Reactive power control loop used for renewable DGs. Source: Ellis et al. 2012 [34].

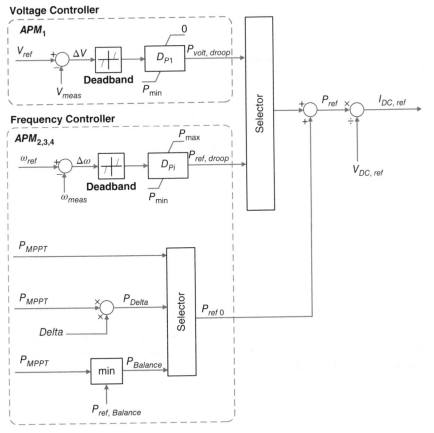

**Figure 5.10** Active power control loop used for renewable DGs. Source: Ellis et al. 2012 [34].

**Table 5.3** Input and output labels description for the electronically coupled DG control loops as shown in Figures 5.9 and 5.10.

| Input label | Description |
|---|---|
| $RPM_1$ (Volt-VAR mode) | Voltage control – isochronous mode |
| $V_{ref}$ | Reference voltage for $RPM_1$ |
| $RPM_2$ (Volt-VAR mode) | Fixed power factor function |
| $PF$ | Power factor for $RPM_2$ |
| $RPM_3$ (Volt-VAR mode) | Fixed $Q$ and voltage droop |
| $Q_{ref}$ | Reference $Q$ for $RPM_3$ |
| $D_Q$ | Droop voltage gain for $RPM_3$ |
| $APM_1$ (Volt-Watt mode) | Voltage droop |
| $D_{P1}$ | Droop voltage gain for $APM_1$ |
| $APM_2$ (Freq-Watt mode) | Delta control with frequency droop |
| $Delta$ | Percentage margin for $APM_2$ |
| $D_{P2}$ | Droop frequency gain for $APM_2$ |
| $APM_3$ (Freq-Watt mode) | Balance control with frequency droop |
| $Balance$ | Maximum power output for $APM_3$ |
| $D_{P3}$ | Droop frequency gain for $APM_3$ |
| $APM_4$ (Freq-Watt mode) | MPPT with frequency droop (no contribution for under frequency scenario) |
| $D_{P4}$ | Droop frequency gain for $APM_4$ |

| Output label | Description |
|---|---|
| $P$ | Active power output at DG terminals |
| $Q$ | Reactive power output at DG terminals |

**Table 5.4** Voltage regulation control loops descriptions for ESS as shown in Figure 5.11.

| Power control loop | Description | Parameters |
|---|---|---|
| Reactive power (Volt-VAR mode) | Voltage control – isochronous mode | $V_{ref}$ |
| | Fixed power factor function | $PF$ |
| | Fixed $Q$ and voltage droop | $Q_{ref}$, $D_Q$ |

## 5.4.2 Frequency Regulation

Additional primary control loops can be incorporated to regulate the system frequency. These frequency control loops operate typically in three modes: (i) isochronous mode – whereby the energy storage system (ESS) sets the system frequency (current

**Table 5.5** Frequency regulation control loops descriptions for ESS as shown in Figure 5.12.

| Power control loop | Description | Parameters |
|---|---|---|
| Active power (Freq-Watt mode) | $P_{ref}$ with frequency droop | $P_{ref}$, $D_P$ |
| | Frequency control – isochronous mode (current controlled VSI) | $f_{ref}$ |
| | Frequency control – isochronous mode (voltage controlled VSI) | $V_{ref}$, $f_{ref}$ |

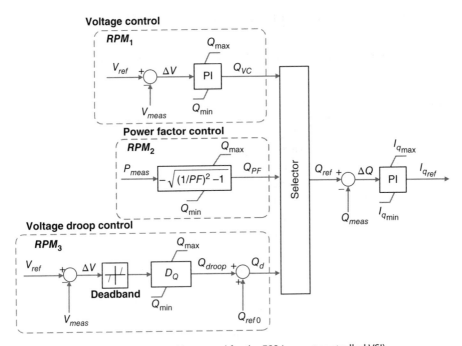

**Figure 5.11** Reactive power control loop used for the ESS (current controlled VSI).

controlled voltage source inverter, VSI) (ii) isochronous mode – whereby the ESS sets the system frequency (voltage controlled VSI) and (iii) droop frequency mode – whereby the ESS is dispatched a power reference $P_{ref0}$ and a droop frequency gain (current controlled VSI). The description for each control loop for frequency regulation is summarized in Table 5.5.

The control loops for the reactive and active power for an electrical ESS are shown in Figures 5.11 and 5.12 respectively. The selector is used to select the operating mode of the ESS. The input and output labels used in the control loops are listed in Table 5.6.

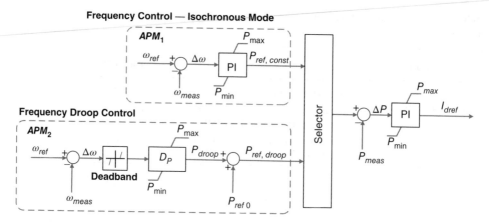

**Figure 5.12** Active power control loop used for the ESS (current controlled VSI).

**Table 5.6** Input and output labels description for the ESS control loops as shown in Figures 5.11 and 5.12.

| Input label | Description |
| --- | --- |
| $RPM_1$ (Volt-VAR mode) | Voltage control – isochronous mode |
| $V_{ref}$ | Reference voltage for $RPM_1$ |
| $RPM_2$ (Volt-VAR mode) | Fixed power factor function |
| $PF$ | Power factor for $RPM_2$ |
| $RPM_3$ (Volt-VAR mode) | Fixed $Q$ and voltage droop |
| $Q_{ref}$ | Reference $Q$ for $RPM_3$ |
| $D_Q$ | Droop voltage gain for $RPM_3$ |
| $APM_1$ (Freq-Watt mode) | $P_{ref}$ with frequency droop |
| $P_{ref}$ | Active power reference for $APM_1$ |
| $D_P$ | Droop frequency gain for $APM_1$ |
| $APM_2$ (Freq-Watt mode) | Frequency control – isochronous mode (current controlled VSI) |
| $f_{ref}$ | Reference frequency for $APM_2$ |

| Output label | Description |
| --- | --- |
| $P$ | Active power output at DG terminals |
| $Q$ | Reactive power output at DG terminals |

## 5.5 Synchronous Generator (SG) Control Loops – Level 1 Function

### 5.5.1 Voltage Regulation

Using reactive power, volt-VAR control can be employed with (i) voltage control – whereby the Synchronous Generator (SG) operates in isochronous mode and set the system voltage, (ii) power factor control – whereby the SG operates at a fixed power factor, and (iii) fixed Q with droop voltage – whereby the SG is dispatched a reference Q and droop gain to regulate the voltage. The IEEE type 1 synchronous machine excitation system is employed. The description for each control loop for voltage regulation is summarized in Table 5.7.

### 5.5.2 Frequency Regulation

Additional primary control loops can be incorporated to regulate the system frequency. These frequency control loops operate typically in two modes: (i) isochronous mode – whereby the SG sets the system frequency and (ii) droop frequency mode – whereby the SG is dispatched a power reference $P_{ref0}$ and a droop frequency gain. The description for each control loop for frequency regulation is summarized in Table 5.8.

The control loops for the reactive and active power for SG are shown in Figures 5.13 and 5.14 respectively. The selector is used to select the operating mode of the SG. The input and output labels used in the control loops are listed in Table 5.9.

## 5.6 Control of Multiple Source Microgrid – Level 1 Function

Another approach toward controlling multiple EC-DERs in a microgrid is presented in this section [36]. The control approach is designed to operate in grid-connected and islanded modes of operation, as well as provide a smooth transition between the two

**Table 5.7** Voltage regulation control loops descriptions for the SG as shown in Figure 5.13.

| Power control loop | Description | Parameters |
|---|---|---|
| Reactive power (Volt-VAR mode) | Voltage control – isochronous mode | $V_{ref}$ |
| | Fixed power factor function | $PF$ |
| | Fixed $Q$ and voltage droop | $Q_{ref}, D_Q$ |

**Table 5.8** Frequency regulation control loops descriptions for the SG as shown in Figure 5.14.

| Power control loop | Description | Parameters |
|---|---|---|
| Active power (Freq-Watt mode) | $P_{ref}$ with frequency droop | $P_{ref}, D_P$ |
| | Frequency control – isochronous mode | $f_{ref}$ |

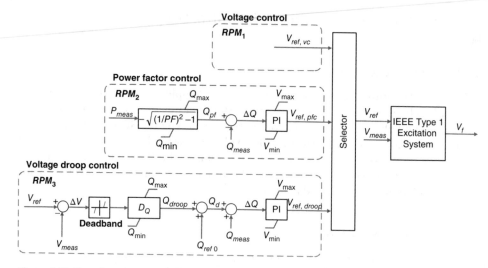

**Figure 5.13** Reactive power control loop used for the synchronous generator [35].

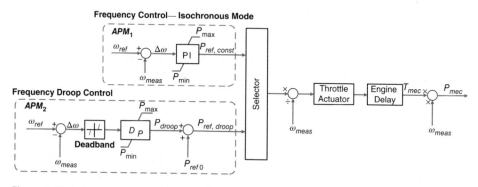

**Figure 5.14** Active power control loop used for the synchronous generator with diesel engine as shown in Figure 3.17.

modes. Additional features including islanding detection with positive feedback and dynamic overcurrent limiting.

The frequency control loop and the voltage control loop for grid-connected, islanded and the transition between the two are given by Equations (5.4) and (5.5) respectively.

$$\frac{1}{\omega_0 K_i}\frac{d\omega_s}{dt} = P_{ref} - P + \frac{1}{\omega_0 D_p}(\omega_{ref} - \omega_p) + \frac{K_p}{\omega_0}(\omega_p - \omega_s) \tag{5.4}$$

$$v_t = V_{ref} - D_q Q + \frac{K_q}{s}(Q_{ref} - Q) + \frac{K_q}{s}G(s)|v_{pcc}| \tag{5.5}$$

where, $\omega_0$, $\omega_s$, $\omega_{ref}$, and $\omega_p$ correspond to the system nominal, converter terminal, point of coupling, and reference frequency respectively, $P_{ref}$ and $P$ are the reference and the measured active power respectively, $Q_{ref}$ and $Q$ are the reference and the measured

**Table 5.9** Input and output labels description for the SG control loops as shown in Figures 5.13 and 5.14.

| Input label | Description |
| --- | --- |
| $RPM_1$ (Volt-VAR mode) | Voltage control – isochronous mode |
| $V_{ref}$ | Reference voltage for $RPM_1$ |
| $RPM_2$ (Volt-VAR mode) | Fixed power factor function |
| $PF$ | Power factor for $RPM_2$ |
| $RPM_3$ (Volt-VAR mode) | Fixed Q and voltage droop |
| $Q_{ref}$ | Reference Q for $RPM_3$ |
| $D_Q$ | Droop voltage gain for $RPM_3$ |
| $APM_1$ (Freq-Watt mode) | $P_{ref}$ with frequency droop |
| $P_{ref}$ | Active power reference for $APM_1$ |
| $D_{P1}$ | Droop frequency gain for $APM_1$ |
| $APM_2$ (Freq-Watt mode) | Frequency control – isochronous mode |
| $f_{ref}$ | Reference frequency for $APM_2$ |

| Output label | Description |
| --- | --- |
| $P$ | Active power output at DG terminals |
| $Q$ | Reactive power output at DG terminals |

reactive power respectively, $K_q$, $K_i$, and $K_p$ are the reactive power integral, active power integral, and active power proportional gains respectively, $D_Q$ and $D_P$ are the reactive and active power droop gains respectively, $V_{ref}$ is the voltage setpoint while $v_t$ is the output terminal voltage.

While the over current is limited by the functions represented by

$$v_{tref} = \begin{cases} |v_{pcc}| - V_{lim} & \text{for } v_t \le |v_{pcc}| - V_{lim} \\ E_s - D_q Q + \frac{K_q}{s}(Q_{ref} - Q) & \text{for } |v_{pcc}| - V_{lim} < e^p_{mag} < |v_{pcc}| + V_{lim} \\ |v_{pcc}| + V_{lim} & \text{for } v_t \ge |v_{pcc}| + V_{lim} \end{cases} \quad (5.6)$$

$$\theta_{VSC} = \begin{cases} \theta_{pcc} - \delta_{lim} & \text{for } \theta_s \le \theta_{pcc} - \delta_{lim} \\ \frac{1}{s}\omega_s & \text{for } \theta_{pcc} - \delta_{lim} < \theta_s < \theta_{pcc} + \delta_{lim} \\ \theta_{pcc} + \delta_{lim} & \text{for } \theta_s \ge \theta_{pcc} + \delta_{lim} \end{cases} \quad (5.7)$$

where, $v_{tref}$ is the commanded output voltage, $v_{pcc}$ is the point of common coupling (PCC) voltage, $\delta$ is the voltage angle of the DER with respect to the PCC voltage angle, $V_{lim}$ is and $\delta_{lim}$ are the hard limits on the PCC voltage and $\delta$ respectively, $\theta_{VSC}$, $\theta_{pcc}$ and $\theta_s$, are the DER output angle, PCC angle, and the integral of the source frequency respectively.

## 5.7 Fault Current Limiting Control Strategy – Level 1 Function

The fault controller supplements the voltage-sag compensation control of a conventional dynamic voltage restorer (DVR). It does not require phase-locked loop and independently controls the magnitude and phase angle of the injected voltage for each phase [37]. The control scheme is shown on a per phase basis in Figure 5.15. The control scheme:

- Can limit the fault current to less than the nominal load current and restore the PCC voltage within 10 ms;
- Can interrupt the fault current in less than two cycles;
- Limits the DC-link voltage rise and, thus, has no restrictions on the duration of fault current interruption;
- Performs satisfactorily even under arcing fault conditions; and
- Can interrupt the fault current under low DC-link voltage conditions.

## 5.8 Mitigating the Impact on Protection System – Level 1 Function

A strategy to mitigate the impact of EC-DERs on the protections systems of a microgrid are discussed in this section [38]. The strategy was developed after studying various fault conditions with different fault resistances and the effects of different DG locations, the impacts of EC-DERs on fuse-recloser coordination in the fuse-saving protection scheme, and the effects of DG reactive power injection, known as a DER potential ancillary service on the protection scheme. The strategy limits the DER output current according to its terminal voltage. In comparison to other methods, this strategy is inexpensive, easy to implement, does not limit DG capacity during normal condition, and does not require any change in the original protection system.

$$I_{ref} = \begin{cases} \dfrac{P_{desired}}{V_{pcc}} & \text{for } V_{pcc} \geq 0.88 \text{p.u.} \\[2ex] kV_{pcc}^n I_{max} & \text{for } V_{pcc} < 0.88 \text{p.u.} \end{cases} \qquad (5.8)$$

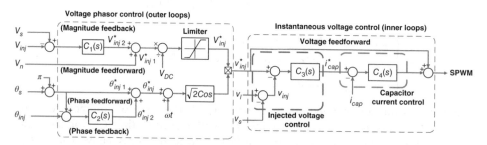

**Figure 5.15** Per-phase block diagram of the DVR control system in FCI mode. Source: Ajaei et al. 2013 [37]. Reproduced with permission of IEEE.

where $I_{ref}$ is the converter reference current, $I_{max}$ is the maximum output current that happens at $V_{pcc} = 0.88$ p.u., $V_{pcc}$ is the rms voltage at the DG connection node, $P_{desired}$ is the output desired power and $k$ and $n$ are constants to be determined. The value of $n$ determines the sensitivity of the control scheme to a voltage change. A larger value of $n$ leads to more obvious output current reduction with a voltage sag. However, too large a value of $n$ will cause the control scheme to become over sensitive to even a small voltage disturbance. Once the value of $n$ is chosen, the coefficient $k$ can be determined in such a way that the reference current in (5.8) remains to be a continuous function around $V_{pcc} = 0.88$; i.e. can be obtained from (5.9).

$$k \, (0.88)^n \, I_{max} = \frac{P_{desired}}{0.88} \Rightarrow k = \frac{P_{desired}}{(0.88)^{n+1} \, I_{max}} \tag{5.9}$$

## 5.9 Adaptive Control Strategy – Level 2 Function

This section presents an adaptive control strategy to augment the existing controllers and enhance their performance [39]. It allows Set Point Automatic Adjustment with Correction Enabled (SPAACE), as an add-on strategy, to improve set point following of power system devices, especially those integrated in a small system such as a micro-grid where the system parameters may change frequently. The strategy monitors the trend and instantaneous values of the response of a device and modulates its set point to achieve a response with a small settling time and small excursion from the set point, e.g. a small overshoot.

The overall system diagram illustrating the placement of the SPAACE controller in conventional power system device is shown in Figure 5.16 and a finite state machine representation of the SPAACE algorithm is shown in Figure 5.17.

## 5.10 Generalized Control Strategy – Level 2 Function

This section presents a control and management strategy for the integration of DERs in microgrids [40]. A hierarchical framework for the control of microgrids – the building blocks of the smart grid – is conceptualized and the notion of poten-tial functions for the secondary controller for devising intermediate set points to ensure feasibility of operation is developed. A potential function is defined for each controllable unit of the microgrid such that minimizing the potential function

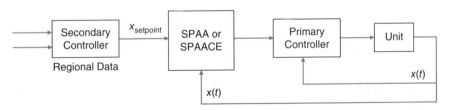

**Figure 5.16** Primary and secondary controllers and SPAACE. Source: Mehrizi-Sani et al. 2012 [39] Reproduced with permission of IEEE.

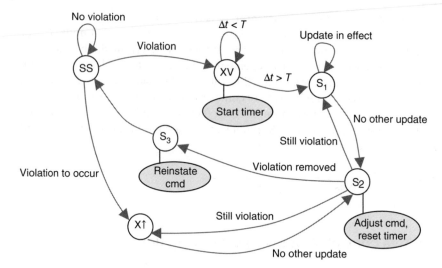

**Figure 5.17** Finite state machine representation of SPAACE algorithm $\Delta t$ is the time passed since violation and $T$ is the maximum permissible time for the violation. Source: Mehrizi-Sani et al. 2012 [39] Reproduced with permission of IEEE.

corresponds to achieving the control goal. The set points are dynamically updated using communication within the microgrid. This strategy is generalized to include both local and system-wide constraints.

Consider the general optimization problem presented by Equation (5.10)

$$\min_x f(x)$$

$$\text{subject to} \quad g(x) = 0$$

$$h(x) \le 0 \tag{5.10}$$

The problem is transformed to Equation (5.11) by adding a barrier function to the objective function.

$$\min_x f(x) - \gamma \sum_{i=1}^{n_i} \log(z_i)$$

$$\text{subject to} \quad g(x) = 0$$

$$h(x) + z = 0$$

$$z \ge 0 \tag{5.11}$$

The problem is then solved iteratively with initial values for $\gamma$, $x$ and $z$ as unity, the current measurements and a vector of random positive numbers respectively and $n_i$ is the number of inequality constraints.

$$\begin{bmatrix} M & g_x \\ g_x^T & 0 \end{bmatrix} \begin{bmatrix} \Delta x \\ \Delta \lambda \end{bmatrix} = - \begin{bmatrix} N \\ g(x) \end{bmatrix} \tag{5.12}$$

$$\Delta \mu = -\mu + [z]^{-1}(\gamma \mathbf{1} - [\mu]\Delta z)$$

$$\Delta z = -h(x) - z - h_x^T \Delta x \tag{5.13}$$

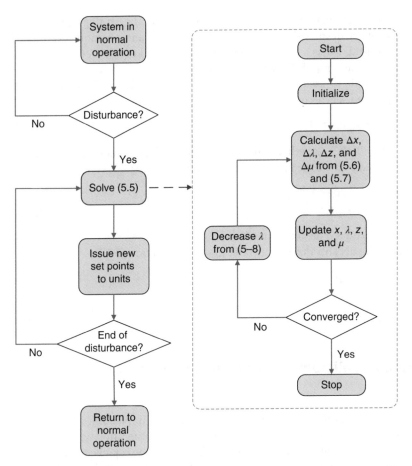

**Figure 5.18** Flowchart of the strategy. Source: Mehrizi-Sani et al. 2012 [40]. Reproduced with permission of IEEE

where, $M$, $N$, and $g_x$ are the coefficients based on the second-order approximation of the Lagrangian, $\lambda$ and $\mu$ are the Lagrangian multipliers for the equality and inequality constraints respectively and $\gamma$ is updated as per Equation (5.14) with $\sigma = 0.1$.

$$\gamma \leftarrow \sigma \frac{z^T \mu}{n_i} \tag{5.14}$$

The flowchart of the solution method is shown in Figure 5.18. This strategy introduces a generalized potential function minimization (GPFM) framework to design the trajectory of the system subsequent to disturbances. GPFM handles both system-wide and local constraints.

## 5.11 Multi-DER Control – Level 2 Function

The mutli-DER control strategy [12, 13] provides a central power-management system (PMS) and a decentralized, robust control strategy for the autonomous mode of operation of a microgrid that includes multiple DER units. The strategy provides a controller that is robust despite the parametric, topological, and un-modeled uncertainties of the microgrid. It has a superior performance in tracking the set-points with zero steady-state error, and rapid disturbance rejection.

The control for the plant (assumed with three sources) provided by Equation (3.14) includes a decentralized servo-compensator given by:

$$\dot{\eta}_i = 0\eta_i + (y_i - y_{ref}^i), i = 1, 2, 3 \tag{5.15}$$

where, $\eta_i \in \mathfrak{R}^2, i = 1, 2, 3$, together with a decentralized stabilizing compensator, which will be assumed to have the structure as in Equation (5.16)

$$\dot{\beta} = A\beta + By$$
$$u = K_1 y + K_2 \eta + K_3 \beta \tag{5.16}$$

A, B, $K_1$, $K_2$, and $K_3$ are $3 \times 3$ diagonal matrices.

This brings us to the decentralized controller given by Equation (5.17)

$$u = K_1 y + K_2 \eta + K_3 \beta$$
$$\dot{\eta} = y - y_{ref}$$
$$\dot{\beta} = A\beta + By \tag{5.17}$$

## 5.12 Centralized Microgrid Controller Functions – Level 3 Function

A preliminary centralized microgrid controller is designed to implement an optimization algorithm in order to determine power set points for the generation units and storage [27, 41]. Whereas the planned microgrid controller should be capable of optimizing more than one objective, only a cost minimization of the energy supplied to the load is considered at this first step. The central controller obtains measurements and device statuses from the different components, performs the optimization, and determines set points for the next time period. In other words, the microgrid controller solves a basic economic dispatch problem allocating the total demand among the different DERs so that the production cost is minimized. The optimization algorithm implemented is the interior point method and the dispatch is performed every 15 minutes. The economic dispatch problem is formulated as,

$$\min \sum_{i=1}^{n_n} C_i(P_{Gi}) \tag{5.18}$$

where $P_{Gi}$ is the generating cost in terms of the amount of power produced by generating unit.

The unit cost functions for the diesel and synchronous generators are quadratic and defined by:

$$C_i(P_{Gi}) = C_{0i} + a_i P_{Gi} + \frac{1}{2} b_i P_{Gi}^2 \tag{5.19}$$

where, $C_{0i}$, $a_i$, and $b_i$ are the quadratic cost parameters.

The constraints considered for the optimization problem are as follows:

1. Power balance is met at all times

$$\sum_{i=1}^{n} P_{Gi} - \sum P_L = 0 \tag{5.20}$$

where, $P_L$ is the total load

2. Every DER must operate within its operating limits

$$P_{Gi,min} \leq P_{Gi} \leq P_{Gi,max} \tag{5.21}$$

Additional constraints such as reserve requirement, ramping limits, DERs turn on/off logic, capacity and energy bounds should all be included in forthcoming studies.

## 5.13  Protection and Control Requirements

Details of a recloser-fuse coordination scheme for improved protection of a microgrid [42, 43] are presented in this section. The protection scheme is effective in both operational modes of a microgrid (i.e. the grid-connected mode and the islanded mode), considering the system constraints; the protection strategies provide the microgrid with acceptable reliability against different fault scenarios.

The protection algorithm is presented in Figure 5.19. The following steps need to be followed to generate the inputs for the algorithm presented:

1. Check whether the coordination of the recloser and the fuse is lost due to the introduction of the DER(s). If the coordination is lost, the algorithm is followed; otherwise, it is stopped in this step.
2. Considering the presence of DERs, the minimum and the maximum fault current of the feeder (i.e. $I_{fmin}$ and $I_{fmax}$) are calculated.
3. The characteristic curves of the devices are coordinated based on the conventional protection scheme.
4. Since, in the presence of DERs, the recloser fault current $I_R$ is less than the fuse fault current $I_F$, the re-closer fast characteristic curve (obtained in Step 3) is revised, through its multiplication by the minimum value of $I_R/I_F$. The revised characteristic curve is then programmed in the digital recloser. In case it is necessary, the instantaneous reclosures are also allowed to ensure a "fuse-saving" mechanism.

## 5.14  Communication-Assisted Protection and Control

This section explains a communication-assisted protection strategy implementable by commercially available microprocessor-based relays for the protection of medium-voltage microgrids [44]. Even though the developed protection strategy benefits from communications, it offers a backup protection strategy to manage communication network failures. The strategy does not require adaptive protective devices; more importantly, its effectiveness is to a large extent independent of the type, size, and location of the microgrid distributed generators; fault current levels; and the microgrid operational mode.

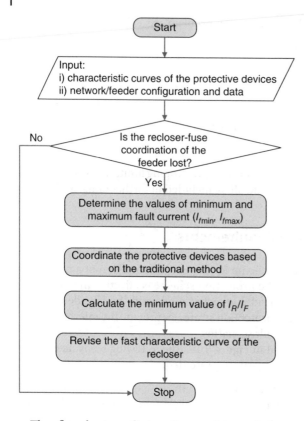

**Figure 5.19** The recloser-fuse coordination algorithm. Source: Zamani et al. 2012 [44] Reproduced with permission of IEEE..

The flowchart outlining the working of the strategy is shown in Figure 5.20. In the figure, CMPR – Communication-Assisted Microgrid Protection Relay, MPC – Microgrid Protection Commander, and CB – Circuit Breaker.

## 5.15 Fault Current Control of DER

This section presents two add-on features for the voltage-control scheme of directly voltage-controlled distributed energy resource units (VC-DERs) of an islanded microgrid to provide overcurrent and overload protection [45]. The overcurrent protection scheme detects the fault, limits the output current magnitude of the DER unit, and restores the microgrid to its normal operating conditions subsequent to fault clearance. The overload protection scheme limits the output power of the VC-DER unit. The flowchart of the scheme is shown in Figure 5.21.

Furthermore, the voltage control strategy is implemented as the following function:

$$\Delta V = \begin{cases} \Delta V_{max} & \text{for } \Delta V > |\Delta V_{max}| \\ \Delta V & \text{for } \Delta V \le |\Delta V_{max}| \end{cases} \tag{5.22}$$

where, $\Delta V_{max}$ is the maximum voltage difference corresponding to the maximum apparent power that the EC-DER can produce while $\Delta V$ is the voltage produced by the EC-DER.

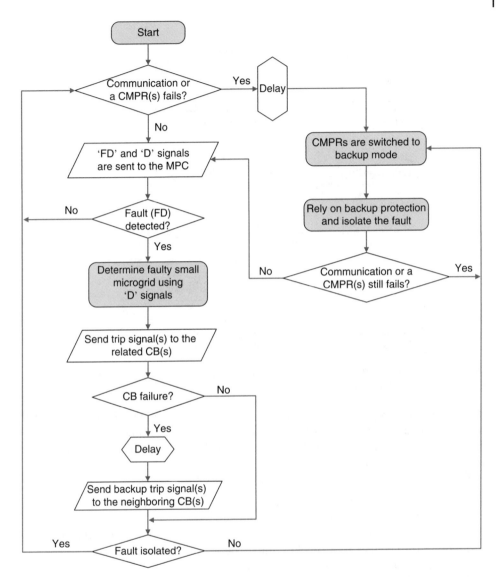

**Figure 5.20** Flowchart of the communications-assisted protection scheme. Source: Etemadi et al. 2013 [45] Reproduced with permission of IEEE..

## 5.16 Load Monitoring for Microgrid Control – Level 3 Function

This section presents a method to monitor loads using a set of distributed voltage sensors and current estimation algorithms [46]. The technique is especially suited for systems where it is difficult to measure loads directly but the load voltages can be sensed, like microgrids. A state-estimation-like current estimation algorithm is proposed to estimate the load currents using the voltages sensed at the terminals of the

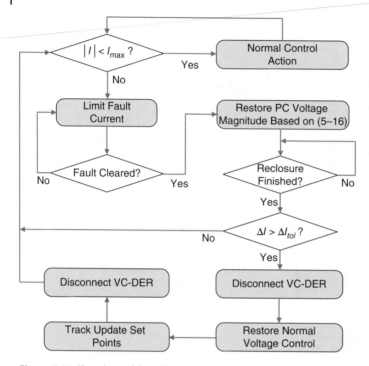

**Figure 5.21** Flowchart of the voltage control scheme including overcurrent protection. Source: Etemadi et al. 2013 [45]. Reproduced with permission of IEEE.

loads. Currents and powers collected from a limited number of panels are also utilized to provide redundancy for the estimation. A major feature of the microgrid is its diverse load management opportunities for customers, which calls for innovative techniques for load monitoring.

The load monitoring algorithm has the following steps:

1. Define the system configuration and parameters: Obtain the system configuration, the branch impedances, and the measurements.
2. Initialization: Calculate the first estimate of the load currents, based on the measurements of the voltages at the load terminals and the main panel voltage.
3. Calculate all branch currents using back sweeping.
4. Calculate all node voltages using forward sweeping.
5. Calculate the change in system state.
6. Update the system state.
7. If is smaller than a convergence tolerance (stop criterion), then stop. If the number of iterations is smaller than the maximum iteration number, go to step 4. Otherwise, conclude that the algorithm does not converge.

## 5.17 Interconnection Transformer Protection

The Restricted Earth Fault (REF) relay is one of the protection units of the transformer digital protection systems which typically has a fast response and thus is prone

**Figure 5.22** Modification in REF logic. Source: Davarpanah et al. 2013 [47] Reproduced with permission of IEEE.

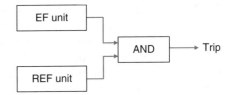

mal-operation. The REF relay should operate only for earth faults (EFs) on the protected winding during the interval that current flow through the neutral current transformer (CT) is inevitable. However, the previous reported studied cases indicate the REF relay can incorrectly operate when there is no significant neutral current.

An EF relay operates when the measured zero-sequence current is more than the pre-set pickup value. The zero-sequence current can be measured based on the sum of the three-phase currents. However, the saturated CT current due to a short circuit can generate an artificial zero-sequence current. Therefore, to determine the actual zero-sequence current value, the current through the transformer neutral should be examined. An EF relay cannot discriminate between the transformer and the system phase-to-ground faults. Therefore, a large delay is often imposed to avoid the relay mal-operation due to external faults.

It must be noted that drawbacks associated with the REF and EF relays can be compensated by the scheme of Figure 5.22 where the trip command of the REF unit is not issued, if the instantaneous EF unit does not operate [47].

## 5.18 Volt-VAR Optimization Control – Level 3 Function

Volt-VAR optimization (VVO) control is a method that optimizes reactive power (VAR) of a distribution network based on predetermined aggregated feeder load profile. This can be done through controlling transformer load tap changers (LTCs), voltage regulators (VRs), capacitor banks (CBs), and other existing Volt-VAR control devices in distribution substations and/or distribution feeders.

Another technology that could be used to potentially reduce the power dispatched to the load in any given feeder is conservation voltage reduction (CVR). Mostly, CVR techniques can conserve energy usage without expecting changes in customer's behavior through maintaining residential voltages on the lower limits of standard range. Additionally, CVR technologies can improve system stability through dynamic control of customer's voltage level. With the advent of real-time command and control capabilities, it is now conceivable to evolve conventional static VVO/CVR systems toward real-time, adaptive, and dynamic VVO/CVR solutions. Furthermore, the availability of dispatchable energy sources in smart grid networks, such as vehicle-to-grid systems, community storage systems, smart inverter technologies, and sustainable/renewable resources, can make VVO/CVR affordable and practical.

The main responsibility of VVO/CVR intelligent agents (IA) shown in Figure 11.11 is to provide an efficient adaptive VVO/CVR management in real time. The central core of this VVO/CVR IA is a volt-VAR Optimization engine (VVOE) that runs real-time optimization and control algorithms in order to maintain the voltage at the point of common connection with the residential consumers within the ANSI C 84.1 standard and CAN

3-C235-83 limits, which is between 0.95 and 1.05 per unit (p.u.), and optimize system voltage, power loss, and active/reactive power simultaneously by applying voltage regulation and VAR injection within the substation and/or along distribution feeders in real time. After reaching the best online optimization solution for the distribution network, it reconfigures the network by sending commands to Volt-VAR assets, e.g. regulation/injection points such as VRs, on-load tap changers (OLTCs), and CBs. These assets implement VVOE's commands in their task zone and if required will send a new command to other IAs in the system. The VVOE platform was developed with the following considerations:

- Maintaining voltage within desired level;
- The importance of distribution network loss;
- Voltage and reactive power control tools;
- VVO; and
- CVR.

# 6

# Information and Communication Systems

This chapter covers the information and communication system requirements for the microgrids. The focus of the systems is on supporting seamless exchange of data and commands between participants in various transactions across the microgrid and the rest of the Electric Power System (EPS). Provisioning and communication issues between various termination points networked as WANs (Wide Area Networks) for substation networking, LANs (Local Area Networks) for smart metering and HANs (Home Area Networks) for smart appliances are explored. It provides a robust communication infrastructure required for the microgrid where the efficient realization of command and control algorithms hinges upon their availability, reliability, and resiliency.

The basic layered model of any communication system is shown in Figure 6.1. Each data unit crosses these layers when being transmitted from one node to another. Microgrids require other dimensional layers as well and are introduced and explained in the following chapter. For communications, each layer is abstracted from the layer above it and handles a specific function to establish, maintain, and/or disconnect the communication session. The application layers handle how the transport layer will be used and for which application. For example, a common application for establishing website sessions is the hyper text transfer protocol (HTTP) while a common protocol used for power system automation and control is the distributed network protocol (DNP3). The transport layer is responsible for controlling how the communication of the data between the two nodes is handled i.e. with guaranteed reliability, error management, etc. Common transport protocols used are transport control protocol (TCP) and user datagram protocol (UDP). The network layer handles how and to where the data is routed. Common network layer protocols include the internet protocol (IP) and the internet control message protocol (ICMP). The data link layer or the network access layer has the job of converting the network layer data to the format that can finally be converted into physical signals that can be sent through the physical media. Common data link layer protocols include the Ethernet, universal serial bus and power line communication (PLC).

## 6.1 IT and Communication Requirements in a Microgrid

The communication systems in a microgrid need to enable an end-to-end connectivity and data exchange between various nodes in the system. Figure 6.2 shows the general network deployments required along with their interfaces. The communication

*Microgrid Planning and Design: A Concise Guide*, First Edition. Hassan Farhangi and Geza Joos.
© 2019 John Wiley & Sons Ltd. Published 2019 by John Wiley & Sons Ltd.

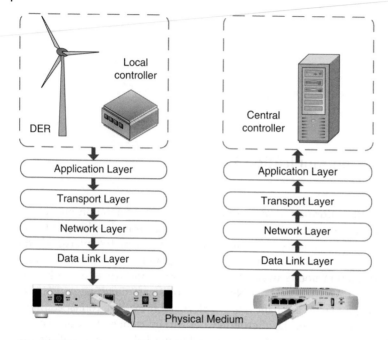

**Figure 6.1** Typical communication layers representation.

networks are classified according to their sizes into WANs, LANs, and HANs and are explained in the following subsections.

### 6.1.1  HAN Communications

HANs provide a network for the non-aggregated atomic nodes devices and enable communications within a home or end-customer environment. A smart meter or other device may provide the gateway between the HAN and LAN. HANs allow devices such as thermostats, appliances, sensors, and in-home displays (IHDs) to communicate with each other in order to make consumers more aware of electrical consumption, and to enable them to control their environment based on variables such as the price of electricity. HANs typically do not require large bandwidth as the volume of network traffic they need to carry is low. OpenHAN and Smart Energy Profile for both ZigBee and Home-Plug are promising standards in future HAN deployment. Furthermore, HANs support applications like energy management systems and demand response technologies at the interface with the utility i.e. the smart meter.

### 6.1.2  LAN Communications

LANs provide a network backbone for the aggregated HANs via gateway applications and protocols like advanced metering infrastructure (AMI) and smart meters. A LAN enables communication in a "neighborhood," where data from several communication-enabled "smart meters," for example, is aggregated. Here a great many different technologies are vying for the attention of utilities. PLC, traditional TCP/IP

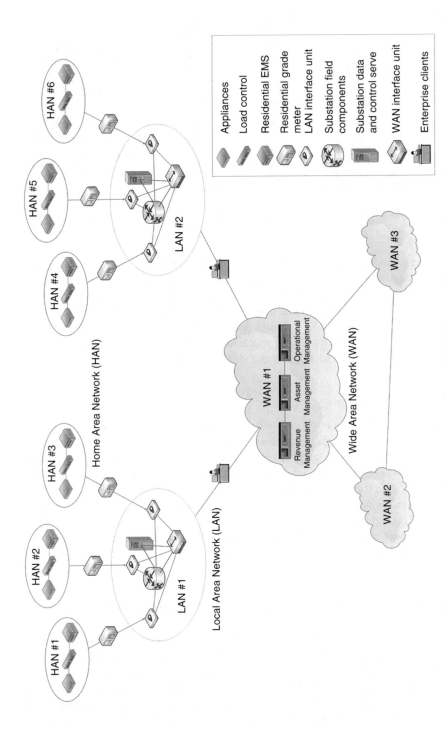

**Figure 6.2** Communication networks overlay between HANs, LANs, and WANs with interfaces applications (inspired from Wu et al. 2012 [48]). Reproduced with permission of CRC.

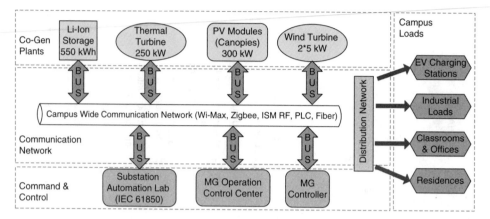

**Figure 6.3** The campus smart microgrid topology and communication network. (Farhangi 2016 [49]).

wired networks, various flavors of wireless ranging from WiMAX to the Industrial Scientific, and Medical (ISM) band RF to ZigBee are all serious contenders in the battle to win this market. Many equipment vendors market proprietary communication solutions where utilities favor a more open solution where interoperability between different vendors' gear is desirable. In North America, ANSI C-12.19 and ANSI C-12.22 are standards that at least address interoperability at the level of the smart meter – a common denominator in the smart grid and smart microgrid arena.

### 6.1.3  WAN Communications

WAN Communications are concerned with backhauling large amounts of data aggregated from several LAN networks, as well as sending command and control information to and from major grid assets such as generation sources, substations, and switching equipment. WANs are usually provisioned with large bandwidth to accommodate the large volumes of data they need to handle. Electrical utilities generally favor traditional TCP/IP networks for WAN communication in conjunction with IEC 61970, IEC 61968, and IEC 61850 for data exchange.

Figure 6.3 shows the campus microgrid communication network as an example of the application of network classifications shown in Figure 6.2.

## 6.2  Technological Options for Communication Systems

This section classifies the available communication technologies into the physical media available for communications and then the technologies that can be implemented on the media. Figure 6.4 summarizes the available options for the communication media. Various communication technologies can be employed for the previously introduced HANs, LANs, and WANs. Several of the most common and applicable technologies for microgrid applications are discussed in this section with their features, advantages, and disadvantages.

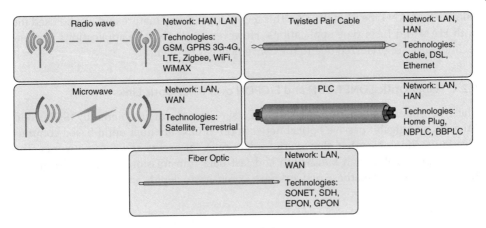

**Figure 6.4** Typical physical media for communication links.

### 6.2.1 Cellular/Radio Frequency

Cellular and radio frequency (RF) based communication technologies like the Global System for Mobile Communications (GSM), General Packet Radio Services (GPRS) 3G and 4G and Long Term Evolution (LTE) networks are primarily used for voice and data connections like the mobile phone and data services. These technologies can also be used for remote monitoring and control for end nodes like DERs and substations. Furthermore, they can also support low frequency small data via the short messaging service (SMS). The technology already has a significant infrastructure installed in most of the countries world wide with wide area coverage. However, if network coverage expansion is required for the microgrid, the investment will be difficult to justify economically if the sole user is the microgrid. Since the technology is based on a wireless medium, it is susceptible to noise and interference resulting in reduced data transfer capacity.

### 6.2.2 Cable/DSL

The cable or the digital subscribers line (DSL) provides a communication service on the existing telephone wires over a different frequency range than that used for voice. This communication technology is suitable for LAN applications where the data is aggregated from various HANs or smart meters. The technology will cost lower in implementation as the infrastructure exists, however, the data throughput diminishes with the increase in the wire length for this technology, therefore, many concentrators or repeaters will be required to aggregate the data of a microgrid covering a large area.

### 6.2.3 Ethernet

Ethernet is a wire-based communication technology that conventionally supports 10 Megabits per second (Mbps) with its further variants: fast Ethernet that supports 100 Mbps and gigabit Ethernet that supports 1000 Mbps or 1 Gbps of data speeds. Higher data speeds are also available. The communication technology is most commonly implemented over a set of four twisted pair wires, while fiber-optic versions for very high-speed Ethernet is also available. This communication technology is most suitable for LAN type of networks. This technology is very mature and can be used for both HAN and LAN type applications. However, for HAN type applications wireless technologies are preferred due to the convenience they offer.

### 6.2.4 Fiber Optic SONET/SDH and E/GPON over Fiber Optic Links

The synchronous optical network (SONET), synchronous digital hierarchy (SDH), and Ethernet and gigabit passive optical network (E/GPON) are fiber optic-based communication technologies that allow various data rates for access and core networks. The technologies enable high bandwidth and fast data transmission capacity with minor interference. However, these networks are expensive to deploy and maintain.

### 6.2.5 Microwave

Microwave is a wireless communication technology that requires the transceivers to be in each others line of sight for communication. The microwave link can be between two nodes directly or with a satellite in between. The link is cable of providing a medium for LAN or WAN type of networks, however, they are much more expensive when compared to other LAN or WAN options like the fiber optic links. The microwave links are also susceptible to environmental factors such as weather and have limited penetration capabilities. Cellular companies already use microwave links to connect different base stations to each other and the server.

### 6.2.6 Power Line Communication

PLC is a technology that enables re-use of the existing power network as a communication medium. This is made possible by creating a communication channel at a much higher frequency than the ones that normally occur in a power network. The technology, though not optimal as the power cables are not specifically designed for communication, can be used for broadband applications (Broad band power line communications [BBPLC]) as well as for narrow band applications (Narrow band power line communications [NBPLC]). The technology is complex to design for and expensive to deploy for power networks with high signal attenuation. Home plug is a low bandwidth PLC based technology for HAN applications.

### 6.2.7 WiFi (IEEE 802.11)

WiFi is a wireless communication technology that is most suitable for home area networks. It allows networking of devices located in close proximity with low cost expansion options. The technology is mature and therefore features rapid deployment and high flexibility. The technology is however more sensitive to interference and vulnerable to security attacks.

### 6.2.8 WiMAX (IEEE 802.16)

WiMAX is also a wireless communication technology that supports higher data speeds and is most suitable to be deployed as intra-HAN aggregation network. It supports a wider range than WiFi and also has a faster deployment speed. The technology however consumes high power, has low penetration when operated at higher frequency bands and is also vulnerable to security attacks.

### 6.2.9 ZigBee

ZigBee is an IEEE 802.15.4 based low power wireless communication technology that allows networking of closely located nodes e.g. HANs. The deployment of the technology is also quicker now that numerous ZigBee certified products are commercially available. The technology supports a data rate of up to 250 Kbps making it suitable for control and sensor networks. The ZigBee smart energy profile is an adaptation of the ZigBee standard for smart energy applications like demand response and home energy management.

## 6.3 IT and Communication Design Examples

Four main categories were identified to be focused on when conducting the IT and communication design studies for a microgrid. The categories are discussed in the following subsections.

### 6.3.1 Universal Communication Infrastructure

Designing a universal communication infrastructure for the microgrid consists of the following steps

- Developing interim guidelines for single dwelling units, multiple dwelling units, neighborhood and wide-area environments based upon best available information;
- Conducting wireless measurement campaigns in single and multiple dwelling units;
- Conducting interference studies and developing deployment guidelines for wired and wireless networks in single and multiple dwelling units;
- Conducting measurement campaigns in neighborhoods and wide area environments; and
- Conducting interference studies and developing deployment guidelines for wide area environment.

The outcome of the mentioned exercises helps generate a list of the required IT and communication specifications.

### 6.3.2 Grid Integration Requirements, Standard, Codes, and Regulatory Considerations

Issues of grid integration requirements, standards, codes, and regulatory considerations in intelligent microgrids are discussed in this section. Specifically, it aims to study and develop efficient transmission, information processing, and networking techniques and strategies suitable for a robust communications infrastructure that supports the integration of intelligent microgrids. The process involves the development and evaluation of

integration strategies for transmission, information processing, and networking architectures, based on available and emerging communications technologies.

It is recommended to follow these steps:

– Characterize different information types in intelligent microgrids to establish their
quality-of-service (QoS) parameters and to classify their dynamic QoS requirements;
– Investigate grid integration requirements, and standards, codes, and regulatory issues
of emerging communications systems in supporting intelligent microgrids.

Based on the studies performed on the benchmark systems discussed in Chapter 2 the
following schemes are recommended.

### 6.3.2.1 Recommended Signaling Scheme and Capacity Limit of PLC Under Bernoulli-Gaussian Impulsive Noise

The optimal signaling scheme and capacity of the Bernoulli–Gaussian impulsive noise
channel was studied to shed new light on the impact of impulsive noise on spectral efficiency of power-line communications systems [50]. First, by focusing on the practically
typical case of impulse power that is much higher than signal power, a tight approximation to the differential entropy of Bernoulli–Gaussian noise was developed with the
derivation of closed-form tight lower and upper bounds on the capacity. A comparison
of the bounds shows that the capacity decreases with an increasing impulse occurrence
rate and the Gaussian signaling scheme is nearly optimal. When the impulse power is
lower than the signal power, the Gaussian signaling can approach the capacity in this
region as well. In addition, channel erasure is shown to be very effective for the impulsive noise channel when impulse power is higher than signal power, but it introduces
rate loss when impulse power is sufficiently lower than signal power.

### 6.3.2.2 Studying and Developing Relevant Networking Techniques for an Efficient and Reliable Smart Grid Communication Network (SGCN)

QoS differentiation and robustness the Greedy Perimeter Stateless Routing (GPSR) and
the Routing protocol for low power and lossy networks (RPL) in microgrid communication networks was studied [51]. Since most smart meters and other microgrid communications devices are installed in harsh outdoor environments, they could fail or wireless
links between them could fluctuate over time. These dynamics could hinder the network
connectivity and degrade the reliability of data communications. Therefore, the Proactive Parent Switching (PPS) scheme effectively helps the RPL routing protocol deflect
network traffic from points of failures in microgrid communications networks [52].

## 6.3.3 Distribution Automation

Distributed automation components including sensors, condition monitoring and fault
detection equipment are discussed in this section. The following schemes are recommended to achieve the mentioned goals.

### 6.3.3.1 Apparent Power Signature Based Islanding Detection

This section presents a passive islanding detection method for distributed generation
units based on extracting signatures from the instantaneous three-phase apparent powers determined at the point of common coupling (PCC) [53, 54]. This method is based

on the fact that the instantaneous apparent powers have components continuously exchanged between loads and sources. The islanding condition creates transient high-frequency components in the instantaneous apparent powers. Therefore, these high-frequency components contain signature information capable of identifying the islanding condition. These transient high-frequency components are extracted using the wavelet packet transform (WPT), when applied to the $d$ $q$-axis components of the instantaneous apparent powers.

### 6.3.3.2 ZigBee in Electricity Substations

The ZigBee wireless platform is a cost-efficient wireless networking system used recently to monitor substation components in electric substations. Although this system has some inherent resistance to interference given its spread spectrum technology, impulsive noise with a short duration and a strong energy content caused by partial discharge (PD) of a dielectric breakdown can degrade the communication quality of ZigBee nodes. Evaluation of the impact of the impulsive noise on the ZigBee 2.4 GHz and 915 MHz frequency bands show that for concurrent deployment of ZigBee with telemetry, the 915 MHz frequency band should be the preferred choice [55, 56].

## 6.3.4 Integrated Data Management and Portals

The section discusses the recommended integrated data management strategies in a microgrid.

### 6.3.4.1 The Multi Agent Volt-VAR Optimization (VVO) Engine

A multi agent Volt-VAR optimization (VVO) engine is presented in this section. Multi-Agent Systems (MASs) are distributed networks that contain intelligent hardware and software agents and work together to achieve a global goal. To enable agents to work together to achieve power system objectives, MAS allocates a local view of power system to each agent, empowering a group of agents to control a wide distributed power system. It is recommended that the VVO scheme be implemented in real-time, using protocols such as the IEC 61850 [18]. The data can be gathered by leveraging the AMI [20, 57]. The AMI data can be disaggregated using integer multi-objective optimization of real and reactive power and on-probability matching tuned by appliance dependence rules [58].

# 7

# Power and Communication Systems

This chapter completes the modeling process by integrating the modeling approaches defined in Chapters 2, 3, 4, and 5.The complete microgrid exists as a multidimensional layered structure. The five layers that comprise the model are shown in Figure 7.1. The multidimensional layered structure of the microgrid was derived from the smart grid architecture model [59]. The communication layers defined in Chapter 6 are invoked in each instance of communication occurring in the multilayered model of the microgrid. The overall microgrid modeling approach is shown as a flow chart in Figure 7.2. An example implementation is also presented.

In Figure 7.1, the physical layer or the power layer is the layer where the topology of the microgrid exists. All the connection information of the connected Distributed Energy Resources (DERs), the network parameters like the line impedances and lengths, location of the point of common coupling (PCC), also exist on this layer. The measurement layer contains all the instrumentation that takes place on the physical layer components including the node voltages, the feeder currents, and the network frequency. This layer also includes the estimators and measurement units installed on the physical system to create the instrumentation data. This instrumentation data is passed on to the protection layer that is also the first line of defense of the network against any anomaly that occurs on the network. This layer is responsible for connecting and disconnecting the physical layer components by monitoring and acting on the data it receives from the instrumentation layer. Once the data pass the protection layer they reaches the function layer. This is responsible for the operational control of each device and or multiple devices at the same time. It is also responsible for following the power references that it gets from the layer above or are generated by local controls like droop curves. For example, a control coordination scheme for stable operation of the microgrid would be implemented on this layer. The final and the top layer is the energy management system (EMS) layer. This layer is responsible for economic or optimal operation of the microgrid. This layer communicates and participates in the electric power system (EPS) energy market and makes optimization decisions to ensure the microgrid achieves the goal it was designed for. Examples of applications residing on this layer would be that of unit commitment and economic dispatch.

*Microgrid Planning and Design: A Concise Guide*, First Edition. Hassan Farhangi and Geza Joos.
© 2019 John Wiley & Sons Ltd. Published 2019 by John Wiley & Sons Ltd.

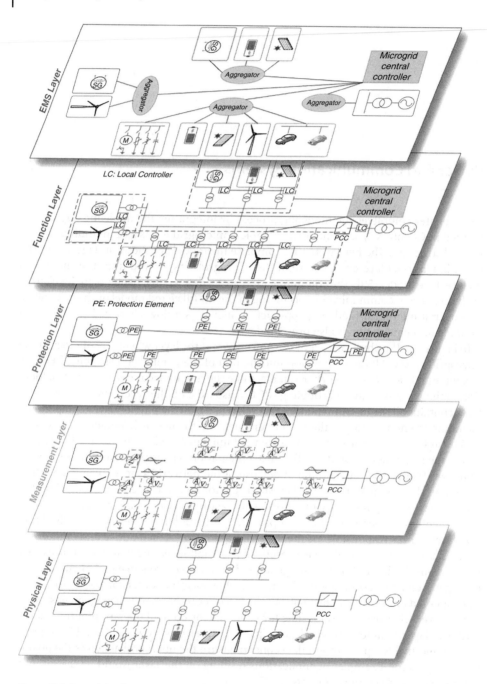

**Figure 7.1** A combined power and communication model.

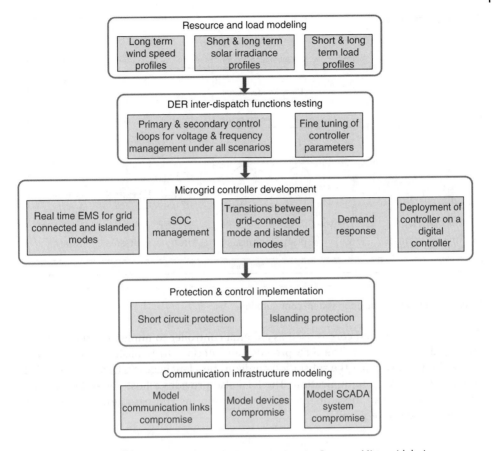

**Figure 7.2** The microgrid modeling approach. Source: Project 2.5 Report – Microgrid design guidelines & use cases – Presented at AGM NSMG-Net Sep. 2015.

## 7.1 Example of Real-Time Systems Using the IEC 61850 Communication Protocol

The IEC 61850 standard, originally put together to ensure interoperability between Intelligent Electronic Devices (IEDs) from multiple manufacturers and simplify commissioning, has gained recent interest in distribution automation due to its real-time, low latency (max 4 ms) generic object-oriented substation event (GOOSE) protocol. The standard allows for communication between devices, where a peer-to-peer model for generic substation events (GSEs) services is used for fast and reliable communication between IEDs. GOOSE allows for the broadcast of multicast messages across a Local Area Network (LAN). The GOOSE messages provide a real-time and reliable data exchange, based on a publisher/subscriber mechanism.

The real-time hardware-in-the-loop (HIL) set-up comprises one real-time simulator (RTS), communicating using the IEC 61850 GOOSE messaging protocol. The overall system is shown in Figure 7.3. One simulator core is used to emulate the network and

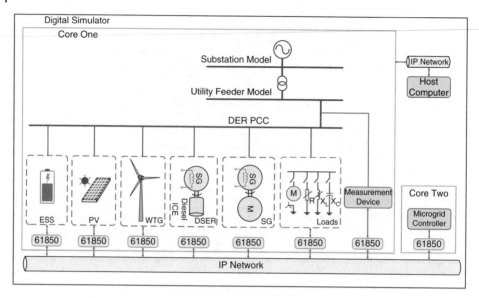

**Figure 7.3** Real time simulation system configuration.

DERs, while the second core is used as a digital controller to implement the proposed microgrid controller [5]. The RTS provides an interface to publish and subscribe to GOOSE messages as per the substation configuration language (SCL) file defining IED configurations and communication. The RTS also provides a host computer interface whereby the user can monitor the real-time waveforms and pass new settings to the real-time model, if required.

# 8

# System Studies and Requirements

This chapter brings together the modeling and validation of various components and control and communication schemes discussed in the previous chapters with their validation and implementation. The chapter provides the data specifications and handling requirements, design criteria, required system studies, and applicable standards to consider for the design methodology presented in Chapter 1 to implement a microgrid.

## 8.1 Data and Specification Requirements

It is imperative to define data structures and requirements for effective communication of the command and control signals throughout the network of actors in a microgrid. Figure 8.1 shows the recommended data and specifications requirements architecture for the simulation and design assessment of the impact of microgrids on the electric power system (EPS).

The main modules shown in Figure 8.1 for the data architecture are explained as follows:

*Module 1. Distribution network model*: In this module, the distribution system is modeled in the simulation program. There will be different microgrid configurations such as mining, remote communities, and campus. Data required for this modeling are: bus and circuit-breaker, transformers, generators, cables, branches, loads, utility switchgear and protective devices. Extensive data on protective devices are also included, such as: fuses, overcurrent relays, reclosers, directional and current transformers.

*Module 2. Microgrid DERs*: Distributed energy resource (DER) penetration can represent a variable within the studies to quantify and qualify the effects of microgrids. However, to conduct studies and analyses with respect to the different power systems, it is important to know about typical DERs currently in operation and foreseen. This requires understanding the following:

 – overall penetration level for the different generation technologies envisaged (Solar Photovoltaic [PV], combined heat and power [CHP], hydro, biomass, wind, etc.) and for different temporal scenarios (for instance: 2018, 2020, 2030);
 – penetration level per interconnection (in MW), for the different generation technologies and scenarios;

*Microgrid Planning and Design: A Concise Guide*, First Edition. Hassan Farhangi and Geza Joos.
© 2019 John Wiley & Sons Ltd. Published 2019 by John Wiley & Sons Ltd.

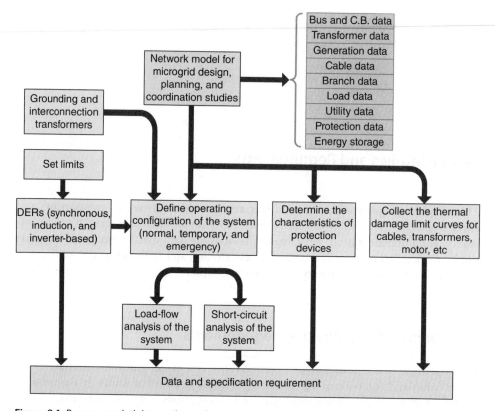

**Figure 8.1** Recommended data and specification requirements for coordinated planning and design.

- distribution of each technology per voltage level (e.g. CHP: 80% LV, 15% MV, 5% HV);
- typical normalized hourly generation profiles for summer, winter, and spring/autumn seasons and for typical weekdays, Saturday, and Sundays.
- Fuel consumption data, assumptions on generation efficiencies and on other characteristics that can be carried out on the basis of average figures for units available on the market.

*Module 3. System operating configuration*: In this module, various operating configuration of the network are defined. These could include normal, temporary, and emergency operating configurations. This module can also be configured by: (i) adding distributed generations (DGs) and (ii) modifying the interconnection transformer vector groups or grounding method.

*Module 4. Protection device characteristics*: In this module, the protective devices characteristics in the system are determined and the time-current characteristics (TCC) are compiled in the simulation program database. The protective device data include: current-tap range, time-dial range, definite time-delay range, and instantaneous trip range.

*Module 5. Equipment thermal limits*: In this module, the thermal damage limit characteristics (i.e. the $I^2t$ curves) for the various equipment in the system including, cables, transformers, and motors are collected.

*Module 6. Grounding and interconnection transformers*: The DGs interconnection transformers could be modeled with the following vector group: Dd (delta-delta), Dyn (delta-star-neutral), Yd (star-delta), Ynd (star-neutral-delta) or Ynyn (star-neutral-star-neutral).

*Module 7. Load-flow analysis*: In this module, the load-flow in the system is computed for every system configuration as defined by module 2 and module 3. The load currents data are used with the equipment continuous ratings to determine the minimum pickup settings of the various protective devices.

*Module 8. Short-circuit analysis*: In this module, the short-circuit currents in the system are computed. The levels of the following currents at each location in the system are computed: (i) Maximum and minimum momentary single- and three-phase short-circuit currents, (ii) Maximum and minimum interrupting duty three-phase short-circuit currents, and (iii) Maximum and minimum ground-fault currents.

In addition to the data required by the aforementioned modules the following characteristic data are also required to perform loss and economic analyses of the microgrid design. The required data are categorized into (i) topology related; (ii) demand related; or (iii) economic/environment related characteristics. These are discussed in the following subsections.

### 8.1.1 Topology-Related Characteristics

Information on the topology of the microgrid and circuits connected are essential. For each voltage level: single-line diagrams, sub-transmission lines, distribution lines, transformers, reclosers, and sectionalizers. In addition, the following data and specification requirements on network characteristics and equipment are:

1. Interconnection Transformers (for each module at each voltage level):
   - rated voltage, power, and load;
   - active power losses;
   - branches impedance;
   - voltage regulation (on-load or off-load tap changer);
   - maximum and minimum voltage variations at all buses;
   - thermal and mechanical damage curves;
2. Feeders:
   - underground (UG) cable, overhead line (OH), and mixed (MX) line;
   - characteristics: cross section (mm$^2$), resistance and reactance ($\Omega$ km$^{-1}$), capacity (MVA);
   - for each circuit in each representative module at each voltage level: cross-sections (mm$^2$); type of circuits (OH/MX/UG [unit commitment]); length (km); percentage and characteristics of loads connected to each circuit; distribution of load connected along the feeder; power factor of the load for each circuit.
3. Network operation:
   - maximum/minimum voltage allowed at the end of the circuit ring, meshed, (open-) loop
   - Losses: average losses (MWh) in different network levels; average fault level (MVA) in different voltage levels.

### 8.1.2 Demand-Related Characteristics

In order to address the actual operational characteristics of microgrid, it is necessary to run simulations based on typical load profiles that resemble the distribution of potential users. Hence, the following load information is required:

- typical user segments (in case depending on the specific country) such as residential (with/without electricity heating), industrial, commercial, and agricultural;
- typical normalized after-diversity load profiles, with one-hour average values for weekday, Saturday, and Sunday in winter, summer, and spring/autumn seasons;
- base load in terms of total peak load in the typical network interconnection;
- overall load distribution among voltage levels and percentage distribution among the different user typologies for each voltage level; and
- expected load increase rate (%/year).

### 8.1.3 Economics- and Environment-Related Characteristics

In order to address the economic and environmental aspects, the following information is required:

- equipment capital costs (cables, lines, transformers, new switchboards);
- electricity charges for active ($/MWh) and reactive ($/MVARh) energy at the different voltage levels;
- charges for reactive power compensation (absorption and generation) for different voltage levels ($/MVARh/year);
- cost of implementation of telecommunication infrastructure needed; and
- environmental analyses including primary energy saving owing to co-generation operation, or $CO_2$ emission reduction owing to active losses reduction in the distribution network

## 8.2 Microgrid Design Criteria

Defining design criteria for the microgrid being planned is very important. The recommended areas to define the design criteria are summarized in Figure 8.2. The following subsections introduce and explain the criteria.

### 8.2.1 Reliability and Resilience

The overall reliability (adequacy and security) of the microgrid and area-EPS, both in effect and planned, exist to ensure that it conforms to the adopted planning criteria while meeting the planning standards. For microgrids, an additional requirement is the resilience, which is the ability of a system to recover itself quickly from a failure or a disturbance. Therefore, the design of a microgrid with respect to the interconnection with area-EPS, restoration devices, and size and type of distributed resources to be installed varies according to the reliability and resilience requirements.

The steps in the following subsections are recommended to increase the reliability and resilience of the microgrid.

**Figure 8.2** Microgrid design criteria.

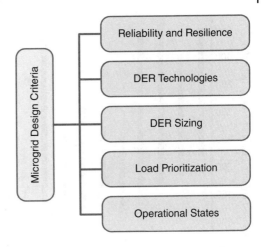

### 8.2.1.1   Reliability

- Provide backup generation resources (DERs or ESS) for critical loads in case of blackout DERs.
- Design and deploy remedial action system such as automatic load shedding scheme to shed noncritical loads under supply shortage.
- Install Microgrid Central Controller (MCC) to coordinate the operations of the microgrid.

### 8.2.1.2   Resilience

- Define interconnection operational rules to secure interconnection with the microgrids under islanding and grid-connected modes.
- Develop tools for restoration prioritization.
- Install intelligent switches technologies, to allow for automatic restoration from alternative power sources.

## 8.2.2   DER Technologies

The deployment of microgrids in residential, commercial, and industrial sectors can have many possible configurations and control structures depending on the available DER technologies and their associated intelligent equipment and control devices. DERs, which include both distributed generation (DG) and energy storage technologies, can be used to secure critical loads during utility outages. DER candidates for a microgrid in campus, for example, may include solar PV, diesel generators, micro-turbines, fuel cells, and energy storage technologies. A comparison of the global levelized cost of energy per megawatt-hour (MWh) of common DG types is shown in Figure 8.3.

The major DER technologies are discussed in the following subsections.

### 8.2.2.1   Electric Storage Systems

Electric energy storage systems can be of various types, but it is only battery and flywheel energy storages have found practical utility in the market due to their rapid response

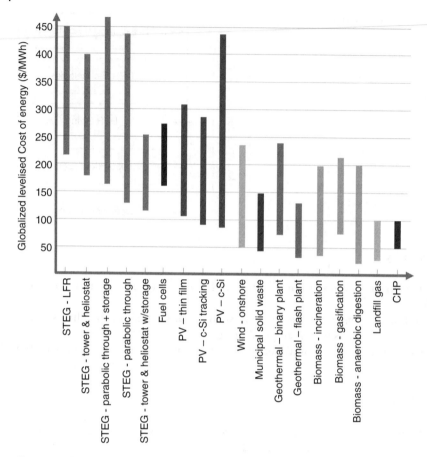

**Figure 8.3** Globalized levelized cost per megawatt-hour comparison of various DER technologies (https://www.worldenergy.org/wp-content/uploads/2013/09/WEC_J1143_CostofTECHNOLOGIES_021013_WEB_Final.pdf - accessed 27th July 2018)

capability. These storage systems can be charged by other DERs in the microgrid or by the area EPS. The following factors have to be considered when planning for ESSs:

- specifying and sizing storage devices, including batteries;
- ensuring safe operation storage batteries;
- life-cycle analysis of storage batteries;
- design of power electronic interfaces; and
- specification and testing of control and protection systems.

A comparison of the estimated levelized cost of energy per kilowatt-hour (kWh) is shown in Figure 8.4 the cheapest storage option is the compressed air energy storage (CAES), however its allowed ramping rate and process efficiency is too low. The most expensive options lithium ion (Li-Ion) batteries, however, they offer much more efficient storage with a high response ramping rate.

### 8.2.2.2 Photovoltaic Solar Power
Solar PV systems extract the energy from the sun and directly convert the solar energy to electricity and provide an opportunity to extract one of the cheapest forms of energy.

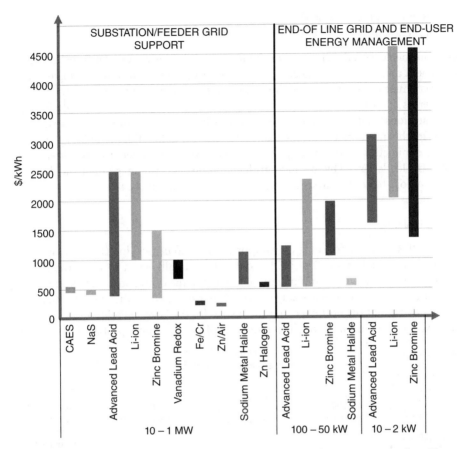

**Figure 8.4** Estimated levelized cost per kilowatt-hour comparison for energy storage (http://www
.eosenergystorage.com/documents/EPRI-Energy-Storage-Webcast-to-Suppliers.pdf - accessed 15
August 2017)

The following factors have to be considered when deciding to install PV systems in a
microgrid:

- Design, layout, and grid connection of solar farms;
- design of power electronic interfaces; and
- specification and testing of control and protection systems.

### 8.2.2.3 Wind Power

Wind turbines systems extract the energy from the wind and convert the kinetic wind
energy via electromagnetic transducers to electric energy. The following factors have to
be considered when deciding to install wind systems in a microgrid:

- reliability, availability, maintainability, and performance assessment;
- AC and DC connection of wind farms to the electric grid;
- monitoring and sensing wind farm production and energy forecasting;
- design of collector systems farms;
- guidelines and performance for energy efficiency design of consumer devices;
- guidelines and performance standards for distribution;

- systems energy efficiency; and
- environmental impact due to the carbon emissions.

Based on the costs, efficiencies, and the footprints of the various DER technologies a cost-benefit analysis can be executed that helps ascertain the best technologies to be used to maximize benefit while meeting the load demand [60]. The benefit can be quantified in terms of cost, environmental well being, energy savings, etc. The analysis helps answer questions related to the economic and technical justification of the creation of a microgrid, the interaction with the main grid, including the utility regulatory requirements for connecting to the main grid, the energy and supply security considerations related to the loads served by the microgrid, including energy management and demand response. A cost-benefit analysis methodology is shown in Figure 8.5.

### 8.2.3 DER Sizing

Sizing the DERs in a microgrid is very important to feed critical loads especially when operating in islanded mode. The DER should be optimally sized to meet the micro-grid design criteria i.e. cost minimization, efficiency maximization, carbon emission minimization, etc. An optimal sizing approach is explained in this section [62]. The approach determines the optimal component sizes for the islanded microgrid, such that the life-cycle cost is minimized while a low loss of power supply probability (LPSP) is

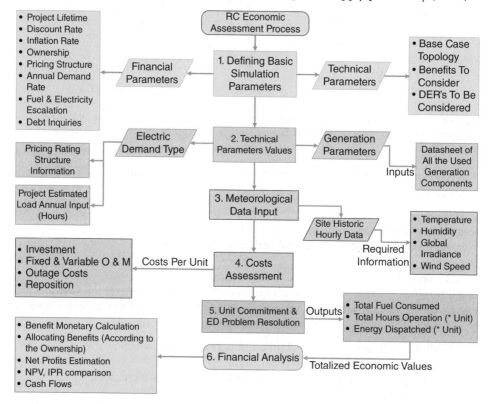

**Figure 8.5** Flowchart of a cost-benefit analysis methodology. Source: Clavier 2013 [61]. Reproduced with permission of McGill University.

ensured. The approach also factors in the diurnal and seasonal variations in the wind speed and solar irradiation profile by utilizing year-based chronological simulation and enumeration-based iterative techniques. The mathematical models used are capable of factoring in the non-linear characteristics as well as the reactive power. The LPSP is formulated based on the supply-demand balances of both real and reactive powers. The sizing approach identifies the global minimum, and simultaneously provides the optimal component sizes as well as the power-management strategies.

The methodology of this approach is shown in Figure 8.6. It starts with specifying the initial value, final value and the step sizes for the elements of the decision variables (the required microgrid components sizes) then applying hourly wind speed either from weather data or from model and hourly solar irradiation and finally the decision

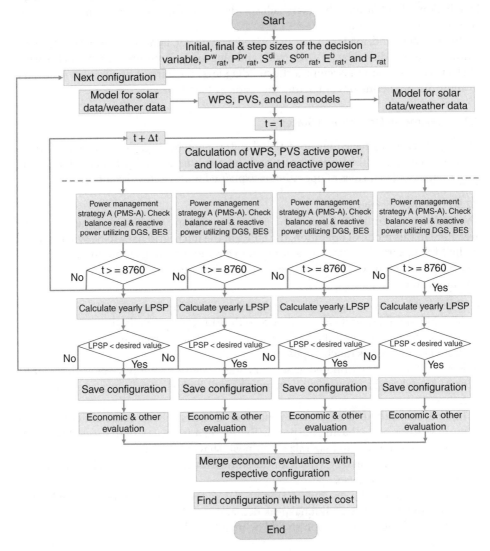

**Figure 8.6** Optimal sizing of islanded microgrids (IMG) methodology. Source: Bhuiyan et al. 2015 [63]. Reproduced with permission of IET.

variables of the optimal sizing model are the rated power of the WPS, $P^w_{rat}$; rated power of the PVS, $P^{pv}_{rat}$ rated power of the charger, $P_{rat}$; rated power of the DGS, $S^{di}_{rat}$; rated power of the battery energy storage system (BESS) converter, $S^{con}_{rat}$; and the rated energy capacity of the battery bank $E^b_{rat}$.

### 8.2.4 Load Prioritization

In microgrids, loads are broadly classified as critical or non-critical loads, and therefore, can be prioritized according to their importance and functions. Consequently, the minimum total generation from DERs should be designed to be just sufficient to serve critical loads. In this case, part of non-critical loads must be designed under load-shedding scheme either through automatic under-frequency or under-voltage. Installation of additional breakers allows disconnecting or connecting the loads according to their priority.

On the other hand, loads can be categorized by type, e.g. lighting loads, heating, ventilation and air conditioning (HVAC) loads and plug loads. Reconfiguration and rewiring of electrical distribution circuits at the building level (i.e. circuit breakers) is necessary to allow disconnecting or connecting the loads by load type.

### 8.2.5 Microgrid Operational States

Microgrids typically have four operational states, namely, (i) grid-connected; (ii) transition to islanding; (iii) islanded mode; and (iv) reconnection to grid. Each state, along with their associated rules of operation is explained in the following subsections.

#### 8.2.5.1 Grid-connected Mode

The grid-connected mode is when the point of common coupling is connected to the area EPS. In this mode of operation, the DERs in the microgrid have to be synchronized to the EPS. In some business cases, for example remote mining communities and military bases, this mode of operation does not exist. The operational rules for this mode of operation are:

- There is no utility outage, and no load shedding should be performed.
- All DERs should operate in accordance with IEEE 1547™ standard.
- For a system with renewable energy sources, electricity from such sources, e.g. PV, should be utilized as much as possible.
- For a system with distributed generators and/or battery energy storage, these DERs can be used to shave the peak demand to avoid high electricity prices during peak hours.
- In most cases, there is a limit on the number of hours a diesel generator can run in a given year for peak-shaving purposes. Therefore, it is necessary to carefully plan for generators' run times to maximize their peak-shaving benefits.
- For storage, it is necessary to plan for the storage unit's charge and discharge schedule to maximize storage use.
- Demand response, for example, changing HVAC temperature set points, can be performed to help reduce additional peak demand.
- Protective device coordination should be maintained.

### 8.2.5.2 Transition to Islanded Mode

Transitions to the islanded mode can be a scheduled event, e.g. initiated by a customer to isolate from the grid during bad weather conditions, or an unscheduled event, e.g. initiated by loss of area voltage or frequency. In an unscheduled event, the key is to sense abnormal conditions from the grid. This can be accomplished by:

- Voltage and frequency sensing;
- Current sensing (magnitude and direction);
- Power flow sensing (magnitude and direction); and
- Others, e.g. phase shift or rate of change of any parameters.

Additional equipment may be added to supplement the functionality of DERs. For example, it may be necessary to dampen any transients produced in the island to prevent a protective relay from tripping-off DERs.

### 8.2.5.3 Islanded Mode

The islanded mode of operation is when the microgrid has successfully disconnected from the area EPS and has undergone any transients that may have occurred right after the disconnection. The following rules should be true when operating in the islanded mode:

- Prior to islanding operation, each DER shall meet the requirements of IEEE 1547™ standard.
- System studies should be performed to support the islanded operation.
- It may be necessary to conduct load-flow and stability studies to identify any potential risks.
- Locally available DERs must be designed to provide the real and reactive power requirements of critical loads.
- They must also provide frequency stability and operate within the voltage ranges as specified in ANSI C84.1.
- Voltage regulation equipment may need to be adjusted to meet the need of the microgrid in the islanded operation.
- Internal DERs should be able to provide adequate reserve margin for the microgrid, considering the load factor, peak load, load shape, reliability requirement, and availability of DERs.
- To balance load and generation:
  - Consideration should be given to achieving generation and load balance for each phase.
  - Load shedding and demand response may become necessary if load requirements are greater than the generation available.
- It may be required to change DER output to match the demand.
- Protective device coordination should be maintained.
- For a system with renewable energy sources, electricity from such sources, e.g. PV, should be utilized as much as possible.
- All other fossil fuel-based DERs should be operated according to their merit orders or as ascertained by the microgrid controller.

#### 8.2.5.4 Transition to Grid-connected Mode

This mode of operation occurs when the area EPS is available again and is ready to accept reconnection of the microgrid. At this point the following rules have to be true:

– Voltage of the main grid must be within the range as specified by ANSI C84.1, before reconnection of the two systems, and frequency range of 59.3 Hz to 60.5 Hz.
– The island reconnection device may delay reconnection for up to five minutes after the system voltage and frequency are restored.
– DERs must be able to adjust the island voltage and frequency to synchronize with the utility grid.

## 8.3 Design Standards and Application Guides

Organizations like the Underwriters Laboratory (UL), American National Standards Institute (ANSI), the Institute of Electrical and Electronics Engineers (IEEE), the International Electrotechnical Commission (IEC), the National Electric Code (NEC) and the International Council on Large Electric Systems (CIGRE) have published a variety of standards, application guides, technical brochures etc. that are applicable at various steps involved in designing a microgrid. The relevant standards for each design procedure are shown in Figure 8.7 and each of the standard is listed as well.

As an example the standards applicable to the modeling of electrical elements of the microgrid.

### 8.3.1 ANSI/NEMA

– ANSI/NEMA C84.1, American National Standard for Electric Power Systems and Equipment – Voltage Ratings (60 Hz).
– ANSI/NEMA MG 1, Motors and Generators.

### 8.3.2 IEEE

– IEEE Std 399™, IEEE Recommended Practice for Industrial and Commercial Power Systems Analysis.
– IEEE Std 446™, IEEE Recommended Practice for Emergency and Standby Power Systems for Industrial and Commercial Applications.
– IEEE Std 519™, IEEE Recommended Practices and Requirements for Harmonic Control in Electrical Power Systems.
– IEEE Std 1100™, IEEE Recommended Practice for Powering and Grounding Electronic Equipment.
– IEEE Std 1547™, IEEE Standard for Interconnecting Distributed Resources with Electric Power Systems.
– IEEE Std 1547.2™, IEEE Application Guide for IEEE Std 1547 Interconnecting Distributed Resources with Electric Power Systems.
– IEEE Std 1547.3™, IEEE Guide for Monitoring, Information Exchange, and Control of Distributed Resources Interconnected with Electric Power Systems.
– IEEE Std 1547.4™, IEEE Guide for Design, Operation, and Integration of Distributed Resources Island Systems with Electric Power Systems.

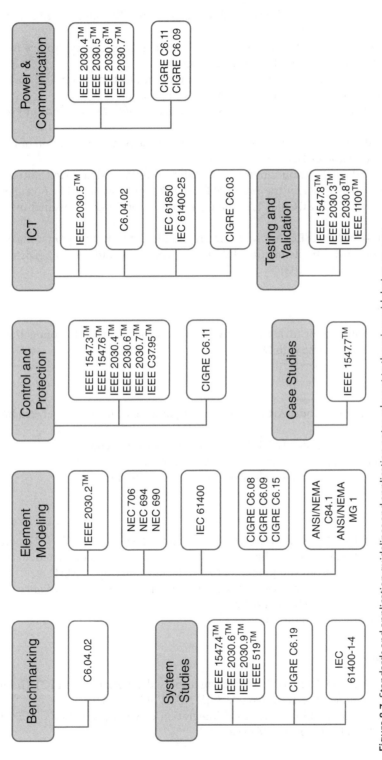

**Figure 8.7** Standards and application guidelines and application notes relevant to the microgrid design process.

- IEEE Std 2030™, Guide for Control and Automation Installations Applied to the Electric Power Infrastructure
- IEEE Std 2030.2 ™, IEEE Guide for the Interoperability of Energy Storage Systems Integrated with the Electric Power Infrastructure
- IEEE Std 2030.3™, IEEE Standard Test Procedures for Electric Energy Storage Equipment and Systems for Electric Power Systems Applications
- IEEE Std 2030.5™, IEEE Adoption of Smart Energy Profile 2.0 Application Protocol Standard
- IEEE Std 2030.6™, IEEE Approved Draft Guide for the Benefit Evaluation of Electric Power Grid Customer Demand Response
- IEEE Std 2030.7™, Distribution Resources Integration WG/Microgrid Controllers
- IEEE Std 2030.8™, WG for the Standard for the Testing of Microgrid Controllers
- IEEE Std C37.95™, IEEE Guide for Protective Relaying of Utility-Consumer Interconnections
- IEEE Std 1815.1™, IEEE Approved Draft Standard for Exchanging Information between networks Implementing IEC 61850 and IEEE Std 1815™, (Distributed Network Protocol – DNP3)

### 8.3.3 UL

- UL1741 – Standard for Inverters, Converters, Controllers, and Interconnection System Equipment for Use with DERs

### 8.3.4 NEC

- National Electrical Code: Article 705 – Interconnected Electric Power Production Sources
- National Electrical Code: Article 706 – Energy Storage Systems
- National Electrical Code: Article 690 – Solar Photovoltaic (PV) Systems
- National Electrical Code: Article 694 – Wind Electric Systems

### 8.3.5 IEC

- IEC/TR 61000-1-4 Electromagnetic compatibility (EMC) – Part 1–4: General – Historical rationale for the limitation of power-frequency conducted harmonic current emissions from equipment, in the frequency range up to 2 kHz
- IEC 60891 Photovoltaic devices – Procedures for temperature and irradiance corrections to measured I-V characteristics
- IEC 61400 series: Wind turbines

### 8.3.6 CIGRE

- Technical Brochure 311, 2007, "Operating dispersed generation with ICT (Information & Communication Technology)," final report of WG C6.03
- Technical Brochure 450, 2011, "Grid integration of wind generation," ELECTRA, February 2011, final report of WG C6.08

- Technical Brochure 457, 2011, "Development and operation of active distribution networks," final report of WG C6.11
- Technical Brochure 458, 2011, "Electric energy storage systems," final report of WG C6.15
- Technical Brochure 475, 2011, "Demand side integration," final report of WG C6.09
- Technical Brochure 575, 2014, "Benchmark systems for network integration of renewable and distributed energy resources," final Report of Task Force C6.04.02
- Technical Brochure 591, 2014, "Planning and optimization methods for active distribution systems," final report of WG C6.19
- Technical Brochure 635, 2015, "Microgrids," first report of WG C6.22

# 9

# Sample Case Studies for Real-Time Operation

System case studies are required for the analysis and evaluation of various microgrid operating scenarios and contingencies, for real time operation (voltage and frequency control, protection, islanding, and reconnection), and energy management of the generation, storage, and loads within the microgrid, in islanded and grid-connected modes. The studies can be categorized into eight different areas and are explained below.

## 9.1 Operational Planning Studies

Operation, planning, and design studies require serious consideration when planning microgrid integration. In the presence of Distributed Energy Resources (DERs) and intermittent renewable energy, this includes optimal power flow, unit commitment, energy management, security, frequency control, and voltage control. New issues that attract increasing attention in the area of operation and control are vehicle-to-grid, demand-side response, and environmental impacts. Each is explained in the following list:

- *Load management and demand-side response.* Quantifying and modeling the benefit and impact of demand-side response.
- *Energy management.* Managing storage for wind energy and photovoltaic (PV) for all modes of microgrid operation including grid-connected, islanded mode, and inter-dispatch operation modes.
- *Voltage control.* Assessing and analyzing the impact of DERs on Volt-VAR control of distribution and transmission including the grid-connected and islanded modes of operation.
- *Frequency control.* Assessing and analyzing the impact of DERs on the power balance in a microgrid and the frequency control including grid-connected, islanded, steady-state, transient, and the emergency load shedding modes of operation.
- *Emissions displacement.* Assessing the value of renewables for emissions displacement.
- *Optimal power flow.* Optimizing losses and cost savings in the presence of DERs.
- *Security.* Self-healing network algorithms.
- *Vehicle-to-grid.* Assessing and analyzing the benefits of hosting electric vehicles on the microgrid capable of offering V2G services including the controls designed for it.

*Microgrid Planning and Design: A Concise Guide*, First Edition. Hassan Farhangi and Geza Joos.
© 2019 John Wiley & Sons Ltd. Published 2019 by John Wiley & Sons Ltd.

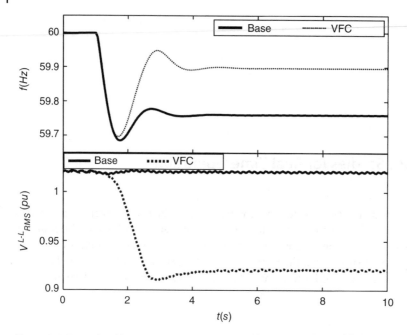

**Figure 9.1** Example of frequency control case study. Source: Farrokhabadi et al. 2015 [64]. Reproduced with permission of IEEE.

- *Black-start and system restoration.* Evaluating and analyzing black-start and system restoration strategies in case of area electric power system (EPS) outages.
- *Protection coordination and selectivity.* Assessing and designing the coordination of protection devices such as re-closers against system faults and incidents such as fault currents (de-sensitization of relays through DERs), fault voltages (voltage support through DERs), grounding (comparing different strategies), and relay tripping (impact of DERs on unwanted tripping).
- *Insulation.* Coordinating and testing for over-voltages that may stress the system insulation.

An example of a frequency control case study is presented in Figure 9.1. The result shows the frequency and the root mean square (RMS) voltage of the CIGRE medium voltage network with several DERs as explained in [64]. The case tested is a sudden disconnection of all the DERs in the system at t = 1 second, which were contributing 2 MW (30%) of the total generation. The base case is the conventional frequency control case while the VFC is the voltage-based frequency control proposed by the authors. This particular result shows the effectiveness of the frequency control proposed. However, the frequency control is achieved by tolerating a lower voltage.

## 9.2 Economic and Technical Feasibility Studies

Economic and technical feasibility studies are done to assess and verify the economic viability of the microgrid project for whichever business case it is being designed. The following components need to be considered when performing these studies:

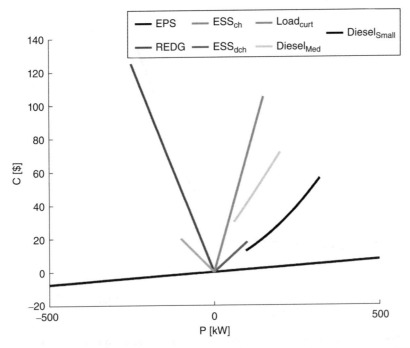

**Figure 9.2** Valuation functions of the individual DERs in the microgrid. Source: Ross 2015 [65]. Reproduced with permission of McGill University.

- *DER sizing.* DER sizing to avoid congestion in the microgrid while meeting the load (including the critical load) requirements of the microgrid.
- *Distribution reinforcement.* Distribution planning for microgrid interconnections.
- *Transmission reinforcement.* Transmission planning for microgrid interconnections.
- *Economic feasibility.* Evaluating the actual economic benefit of the microgrid project including the cost-benefit analysis.

Figures 9.2 and 9.3 illustrate an example of economic feasibility studies. Figure 9.2 shows the valuation functions of the DERs in the case study with linear and negative valuation functions, while Figure 9.3 shows the valuation functions of the amalgamated virtual resources for the same case study. The feasibility study was done to evaluate the operation of individual DERs amalgamated as virtual resources (with DERs assumed to be dispatchable) [65] for the facilitation of optimal economic dispatch and unit commitment problems in a microgrid. The DERs amalgamated as a virtual resource include the area EPS, the renewable energy based Distributed Generation (REDG), energy storage system charging (ESS$_{ch}$) and discharging (ESS$_{dch}$), the curtailed load (Load$_{curt}$), medium diesel generator (Diesel$_{Med}$) and small diesel generator (Diesel$_{Small}$).

## 9.3 Policy and Regulatory Framework Studies

Studies that analyze the impact and conformity of the microgrid with the regulatory policy framework are required to ensure that the microgrid complies with the regulations

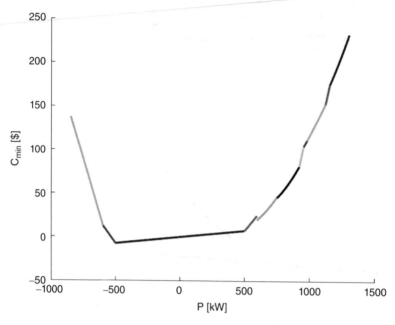

**Figure 9.3** Valuation function of the amalgamated virtual resource for the microgrid at a specific operating point. Source: Ross 2015 [65]. Reproduced with permission of McGill University.

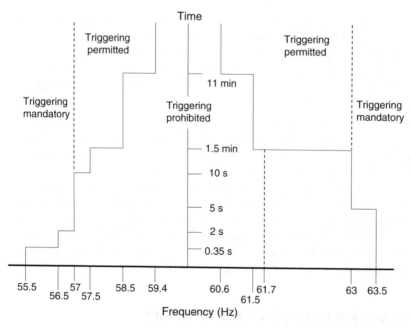

**Figure 9.4** Example of frequency-related utility regulations. Source: http://www.hydroquebec.com/transenergie/fr/commerce/pdf/e1201_fev09.pdf. Reproduced with permission of HydroQuebec.

set by the utility that provides the area EPS. The studies include assessing the regulatory framework, the technical regulatory issues of power systems, and the communications systems. The energy and services market issues should also be considered if the regulations exist. As an example the frequency-related utility regulation curve for DERs connecting to a utility's medium voltage network is shown in Figure 9.4.

## 9.4  Power-Quality Studies

Power-quality studies are important for the microgrid to provide high-quality power to the loads that may be critical (such as hospitals). Apart from that, factors such as harmonics, unbalance, etc. increase system losses and may result in nuisance tripping of the protection relays. Some factors to consider in such studies are:

- *Ferro-resonance*. The presence of DERs in the microgrid or any transient event may excite ferro-resonance (which occurs due to cable capacitance and transformer inductance). Therefore, the risk of ferro-resonance in transformer-connected DERs should be assessed and analyzed.
- *Flicker*. Flicker contributions from distributed wind create a periodic dip in the supply voltage and must be minimized.
- *Harmonics*. Contribution of the DERs to the harmonics in network and ways to mitigate them.
- *Motor*. Providing load starting currents (e.g. motors) with DERs and how it impacts the protection devices and power quality.
- *Service interruption*. Ascertaining system average interruption duration and frequency indices.
- *Unbalance*. Impact of single-phase DER and load connections.
- *Voltage profile*. Reactive power management with DERs.

Results from a harmonics study done on residential loads are shown in Figure 9.5a, b. The figures show that there are various harmonics and subharmonics in the currents absorbed by common household appliances. Harmonics studies are necessary as they add to the reactive power demand and thus reduce the load power factor. This also results in a higher current resulting in greater losses.

## 9.5  Stability Studies

Stability studies provide designers with the points of operation that bring the system closer to, or push it toward instabilities such as voltage collapse, frequency instabilities, low frequency power oscillations, etc. The following factors have to be considered when conducting stability studies:

- *Islanding*. Enhancing stability with controlled islanding.
- *Low-voltage ride through*. Assessing the impact of the low voltage ride through (LVRT) profile on system stability.
- *Small-signal angle stability*. Angle stability in networks with long cables connecting loads and DERs.

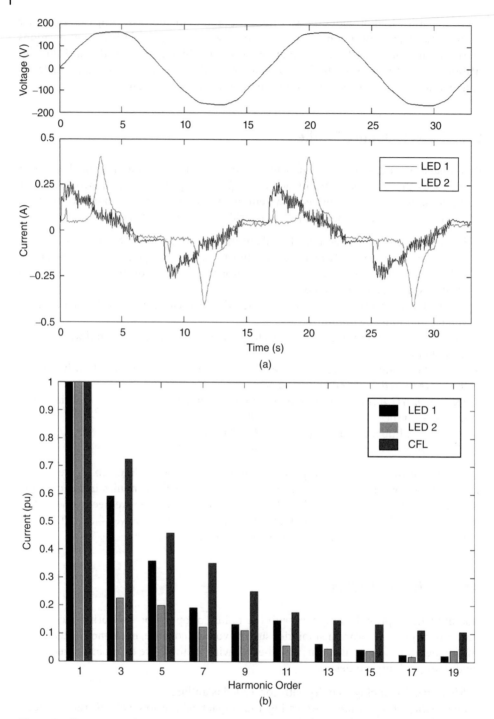

**Figure 9.5** Harmonic study on common household loads. (a) Supply voltage and input current waveforms of two commercial variants of LED lamp; (b) frequency spectrum of commercially available LED and CFL lamps. Source: Wang et al. 2013 [62]. Reproduced with permission of IEEE.

- *Stabilizer design.* Testing stabilizers for distributed generation (DG) in real-time hardware-in-the-loop (HIL).
- *Transient stability.* Assessing the response of DERs subject to large disturbances.
- *Voltage stability.* Voltage collapse assessment.

As an example, results of a stability analysis study with a stabilizer design are shown in Figures 9.6 and 9.7. Figure 9.7 shows the trace of the dominant Eigen values of the CIGRE North American Medium Voltage network as the value of $m$ is increased. In the figure $m$ represents the real power droop gain of an electronically coupled distributed energy resource (EC-DER). Based on the stability analysis a stabilizing scheme for the EC-DER was proposed [33]. Figure 9.6a, b shows the real and reactive power of the

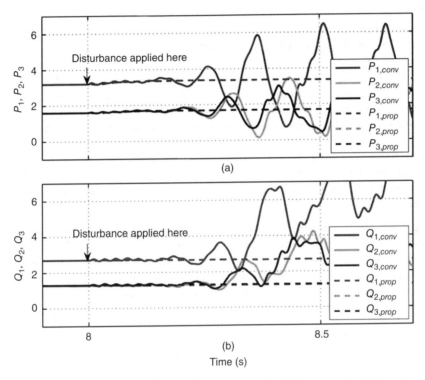

**Figure 9.6** DER response to the proposed droop control and the conventional droop-based control with high gains. Source: Haddadi et al. 2014 [33]. Reproduced with permission of IEEE.

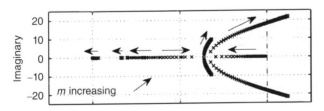

**Figure 9.7** Trace of the dominant eigenvalues of a three DER system for a droop gain. $2 < m_1 < 30$ rad/s/MW, $m_2 = m_3 = 2\,m_1$, where $m_1$, $m_2$, and $m_3$ are the droop gains of $DER_1$, $DER_2$, and $DER_3$ respectively. Source: Haddadi et al. 2014 [33]. Reproduced with permission of IEEE.

DERs after a disturbance was applied. The study shows that the proposed stabilizing control stabilizes the system despite the disturbance, which otherwise would have become unstable. However, the EC-DERs are assumed to have an ideal Direct Current (DC) voltage source on the DC link.

## 9.6 Microgrid Design Studies

Design studies for microgrids involve assessing the technical feasibility of the microgrid in all its operational modes. This involves parameter identification of the actual systems to fit the models. The following studies (several overlapping from other studies) are required:

- *Modeling.* Parameter identification of DERs, the network, and the microgrid control. Ascertaining the modeling, simulations, and software requirements to conduct the studies.
- *Protection, control, and monitoring studies.* Protection planning, designing control strategies, islanding/reconnection management, and compatibility with the existing distribution automation.
- *Energy storage studies.* Sizing and scheduling of energy storage systems in the microgrid.
- *Ancillary services.* Assessing the technical and economic impact of real and reactive power-related ancillary services.
- *Microgrid impact studies.* Impact on the environment; impact on the market; impact on the distribution system.

An example of protection studies is presented here. The synchronous machine generators are known to feed in large currents to faults that result in increased difficulties in protection coordination; especially its overcurrent (OC) protection element. Figure 9.8 shows the RMS alternating current (AC) component of the output current of a synchronous generator when fault occurs, for different values of $k$. The value $k$ is defined as the ratio between the total resistance of the discharge circuit to the field resistance [66].

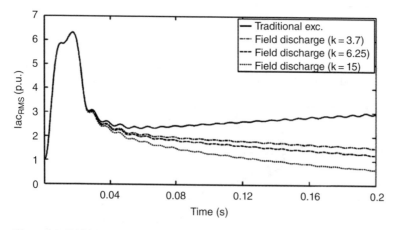

**Figure 9.8** RMS AC component of the output current of the synchronous generator feeding to the fault. Source: Yazdanpanahi et al. 2014 [66]. Reproduced with permission of IEEE.

It can be seen that the output current decreases for all cases except the traditional excitation case. This was exploited by intentionally discharging the field winding during a fault to reduce the generator output current.

## 9.7 Communication and SCADA System Studies

Communications system studies are required to assess and analyze the effectiveness of the employed communications system and its impact on Supervisory Control and Data Acquisition (SCADA) while complying with other design parameters of the microgrid. The following factors are required to be considered for such studies:

- *Communication specifications.* Identifying the requirements and specifications required to support the required SCADA strategies.
- *Grid integration.* Identifying the requirements on the integration of the required communication infrastructure with the microgrid and its impact on its operation and control.
- *Cyber security.* Assessing the impact of cyber attacks on the communication systems and on the operation and control of the microgrid.

As an example, a cyber security study is presented. Figure 9.9 shows the frequency of the system for various cases after a persistent false data injection (FDI) attack on the communication system. The FDI attack is when an attacker falsely injects a control signal, an under frequency load shedding signal in this case, to any grid entity, the ESS in this case [67]. Figure 9.9 shows that the system frequency starts to violate the limits set by standards and regulations if no remedial scheme is applied and it remains within the limits when the remedial schemes proposed by the authors are applied. Despite the remedial scheme the FDI issue needs to be dealt with using cyber security efforts.

## 9.8 Testing and Evaluation Studies

This involves testing and evaluating the complete systems once all the studies are performed to verify, validate, and finalize the microgrid design. The latest available tool for

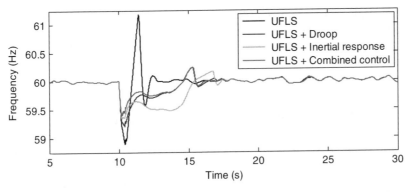

**Figure 9.9** Frequency response subject to a FDI persistent attack. Source: Chlela et al. 2016 [67]. Reproduced with permission of IEEE.

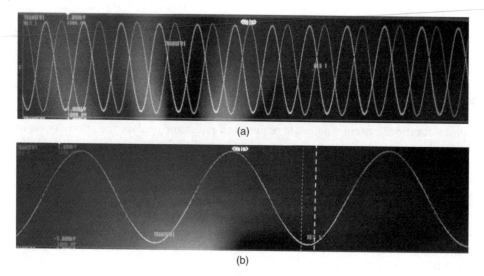

**Figure 9.10** Transition between (a) islanded and (b) grid connected mode of operation of a microgrid. Source: Ross et al. 2014 [5]. Reproduced with permission of CIGRE/HydroQuebec.

real-time testing are the real-time simulators (RTSs) that make it possible to simulate the system and test its control strategies in conditions that are as close to the real world as possible. RTS with HIL capabilities allow testing of the actual commercially available devices such as the relays, energy management systems, etc. with the rest of the microgrid models simulated in the RTS.

A transition between the islanded and grid-connected mode was tested at the utility microgrid and the results can be found in [5]. When signaled to resynchronize, the phase angle difference between the EPS voltage and the microgrid voltage is 150°, as shown in Figure 9.10a. The isochronous generator then modifies its phase angle to synchronize with the grid voltage as shown in Figure 9.10b.

## 9.9 Example Studies

Several studies can be conducted to understand the impacts of integrating different DER technologies on the performance of the microgrid, and verify the robustness of the deployed control strategy in both modes of operation: islanded and grid-connected, as well as the transition between the two modes. A sample of impact studies for grid-connected and islanded mode are:

- *Impacts of sudden load switching during the grid-connected operation.* In this case, the load at any given substation suddenly increases by 100% at a given instant of time. In the grid-connected mode, "sudden" refers to the case when any change in the load or generation is compensated by area EPS instead of being compensated from the battery station or the DER units.
- *Impacts of pre-planned load switching during grid-connected operation.* This case is identical to the previous case except that the load increase is compensated from the battery station to minimize the power exchange with area EPS.

- *Impacts of changing the DER reference set-point.* In this case, the reference set point of the DER connected to any given substation is reduced to half a given instant of time. The resulting power mismatch is compensated from area EPS.
- *Impacts of pre-planned DER outage.* In this case, at a given instant the real and reactive power reference set points of the DER unit at the equivalent substation drop to zero, i.e. the DER is disconnected. However, this DER outage is pre-planned since the corresponding reference power set points of the battery simultaneously increase to compensate for disconnecting the DER.
- *Matched power islanding.* In this scenario, the microgrid islands at a given instant with zero power transfer through the Point of Common Coupling (PCC) breaker prior to islanding, i.e. matched power islanding. Prior to islanding, the PQ controllers of the battery station regulate its output power to zero.
- *Pre-planned unmatched power islanding.* In this scenario, the total system load is 10% larger than the total DER capacity. During the grid connected mode, this power mismatch is compensated by Area-EPS while the real and reactive power set points of the battery station are kept at zero. In this work, "pre-planned islanding" refers to the islanding events with power mismatch less than the battery rating or to prescheduled load shedding to force the power mismatch to be less than the storage capacity prior to islanding.
- *Heavy loading during islanded operation.* In this case, the load at any given substation is suddenly increases by 100% at a given instant. In the islanded mode, "sudden" refers to the case when any change in the load or generation is compensated by the battery station instead of changing the reference power set point of any other DER unit to avoid overloading the battery.
- *Light loading during islanded operation.* In this case, the load at any given substation is suddenly Decreases by 100% at a given instant.

# 10

# Microgrid Use Cases

Constructing use cases is an effective practice to map applications to processes to achieve a certain objective. The procedure involves enlisting the actions or steps that define the interaction between an instigator (or actor) and a process (or system). This creates a simple visual representation of complex systems outlining the requirements. The use cases in this chapter have been developed in terms of use case description, actor roles, information exchange and associations between objects of use case, and regulations. Other use cases from different organizations can be found in [68]. Only the use cases that have been identified as relevant for smart microgrid implementation will be presented here. It covers three publicly available use cases, namely, the energy management system (EMS) functional requirements, protection, and intentional islanding as provided by the Oakridge National Laboratory (ORNL).

## 10.1 Energy Management System Functional Requirements Use Case

This use case defines the EMS functions of a microgrid working in both grid-connected and islanding mode of operations. The microgrid EMS is a part of the microgrid controller (MC), in which it coordinates among multiple distributed energy resources (DERs), storage battery, main grid, and responsive loads to improve system reliability, perform economic dispatch, and reduce the operation cost. In its operation, it manages the power flow, power transaction, energy generation and consumption, real/reactive power, and battery charging/discharging in a microgrid. An EMS can be deployed to achieve the optimization according to day-ahead bidding and scheduling, short-term economic dispatch, and real-time.

In general, a microgrid has two modes of operation, namely, grid-connected and islanding. The operation and system constraints could be different in different modes. In grid-connection, the microgrid EMS coordinates with the area electric power system (AEPS) and manages the microgrid to comply with the AEPS policies, regulations, and requirements, and provides ancillary services accordingly. In islanding mode, the microgrid EMS are to maintain the stability, regulate the voltage and frequency within certain ranges, and optimize the microgrid's overall performances.

*Microgrid Planning and Design: A Concise Guide,* First Edition. Hassan Farhangi and Geza Joos.
© 2019 John Wiley & Sons Ltd. Published 2019 by John Wiley & Sons Ltd.

**Table 10.1** Attributes of the EMS use case.

| Name (diagram label) | Type | Description |
| --- | --- | --- |
| Area Electric Power System (Area EPS) | System | The electrical power system that normally supplies the microgrid through the point of common coupling. |
| Local Electric Power System (Local EPS) | System | The electrical power system on the customer's side of the PCC. |
| Area Fuel Supply (AFS) | System | The fueling system that supplies the microgrid. |
| Point of Common Coupling (PCC) | System | The interface substation between the AEPS and the microgrid. |
| Critical Load (C Load) | Device | The highest priority loads within the microgrid. These loads are not part of the load shedding schemes. |
| Non-Critical Load (NC Load) | Device | The lowest priority loads within the microgrid. These loads may be left unserved in favor of critical loads. |
| Microgrid Controller (MC) | System | A control system that is able to dispatch the microgrid resources including opening/closing circuit breakers, changing control reference points, changing generation levels, and coordinates the sources and loads to maintain system stability. |
| Microgrid SCADA (MG SCADA) | System | Provides the data acquisition and telecommunication required for the microgrid controller functions. It collects real-time data from each microgrid actor, and executes control actions such as economic dispatch commands, circuit breaker controls/status. |
| Primary Distributed Energy Resource (PDER) | Device | The distributed energy resources participating in voltage regulation. PDERs could be a generator and energy storage system. |
| Switching Device (DER-SW) | Device | The DER-SW can disconnect DER within the microgrid. The DER-SW can receive control signals from the MC and can inform the MC of its status through SCADA. |
| Market Operator (MO) | System | The MO accepts bids from MG, in its AEPS and dispatches MG sources to provide energy and ancillary services. The MO may be part of the AEPS or may be a separate entity. |

Source: Reproduced with the permission of ORNL.

The actor roles within the use case are described in terms of the actor attributes, which include name, type, and description in Table 10.1. The information exchange model is described in Table 10.2 and the relevant standards are explained in Table 10.3. The use case diagram is shown in Figure 10.1.

**Table 10.2** Information exchange and associations between objects of the EMS use case.

| Object name | Description |
|---|---|
| Microgrid Measurements and Status (MG Meas. and Status) | Includes voltage, current, frequency, and power (active, reactive) measured at each actor, and the status of the actors, including on/off status, DERs' operation modes (primary DER or as a non-regulating source), as well as other operation status indicators. |
| Microgrid Control Commands (MG Cont. Comm.) | These control commands to microgrid actors (DERs, microgrid EMS, loads, switching devices, protection relays, islanding schemes, and synchronization relays). These commands define each DER's control mode, real and reactive power dispatch, loading for controllable loads, frequency and voltage setting points, islanding, and resynchronization. |
| Microgrid Bid (MG Bid) | Transaction between system operator and microgrid controller. |
| Unit Commitment (Unit Comm.) | DERs planning and scheduling. |
| Economic Dispatch (Eco. Disp.) | Services provided to support the operation of the power network (it could in the form of active or non-active energy). |
| Day-ahead forecast (Day + Forecast) | Generation and demand. |
| Real-time Operation (RT Operation) | Generation and demand. |
| Short-term Forecast (ST Forecast) | Generation and demand. |

Source: Reproduced with the permission of ORNL.

**Table 10.3** Regulations of the EMS use case.

| Regulation | Description |
|---|---|
| IEEE Std 1547™, series | Standard for interconnection to the area EPS at the PCC. |
| Interconnection agreement | Defines interconnection terms and conditions such as: interconnection studies, operations rules, constructions, safety regulations, maintenance policies, access, boundary limits, disconnect circuit breakers/switching/isolation devices, conflicts in agreements, disconnection, customer generator billing and payment, insurance, customer-generator indemnification, limitation of liability, contract termination, permanents, survival rights, assignment/transfer of ownership of the customer-generator facility, telecommunications, tele-control, and tele-metering. |
| ISO market rules | Defines rules when microgrid participates in markets. |
| Metering regulations | Defines regulations when microgrid exports to AEPS. |

Source: Reproduced with the permission of ORNL.

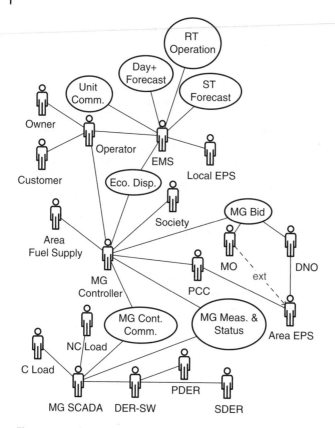

**Figure 10.1** The EMS Functional Requirements use case diagram.

## 10.2 Protection

This use case defines protection systems for different microgrid configurations and operating conditions, in order to detect and isolate faults without impacting the microgrid. A microgrid protection scheme must be designed to prevent personnel risks, minimize equipment damages, and reduce the loss of loads. The microgrid protection controller communicates with the microgrid supervisory control and data acquisition (MG SCADA) periodically to update the system status and determine the protection schemes and settings accordingly. Updated protection settings are distributed through communication links from the protection controller to all the protection devices with communication facilities. Protection relays detect faults and trip circuit breakers either by a local protection device or remotely through communication with the local protection relay. The microgrid topology is updated periodically.

**Table 10.4** Attributes of the Protection use case.

| Name (diagram label) | Type | Description |
|---|---|---|
| Area Electric Power System (Area EPS) | System | The electrical power system that normally supplies the microgrid through the point of common coupling. |
| Local Electric Power System (Local EPS) | System | The electrical power system on the customer's side of the PCC. |
| Area Fuel Supply (AFS) | System | The fueling system that supplies the microgrid. |
| Point of Common Coupling (PCC) | System | The interface substation between the AEPS and the microgrid. |
| Critical Load (C Load) | Device | The highest priority loads within the microgrid. These loads are not part of the load shedding schemes. |
| Non-Critical Load (NC Load) | Device | The lowest priority loads within the microgrid. These loads may be left unserved in favor of critical loads. |
| Microgrid Controller (MC) | System | A control system that is able to dispatch the microgrid resources including opening/closing circuit breakers, changing control reference points, changing generation levels, and coordinates the sources and loads to maintain system stability. |
| Microgrid SCADA (MG SCADA) | System | Provides the data acquisition and telecommunication required for the microgrid controller functions. It collects real-time data from each microgrid actor, and executes control actions such as economic dispatch commands, circuit breaker controls/status. |
| Primary DER (PDER) | Device | The distributed energy resources participating in voltage regulation. PDERs could be a generator and energy storage system. |
| Switching Device (DER-SW) | Device | The DER-SW can disconnect DER within the microgrid. The DER-SW can receive control signals from the MC and can inform the MC of its status through SCADA. |
| Market Operator (MO) | System | The MO accepts bids from MG, in its AEPS and dispatches MG sources to provide energy and ancillary services. The MO may be part of the AEPS or may be a separate entity. |

Source: Reproduced with the permission of ORNL.

The actor roles within the use case are described in terms of the actor attributes which include name, type and description in Table 10.4. The information exchange model is described in Table 10.5 and the relevant standards are explained in Table 10.6. The use case diagram is shown in Figure 10.2.

**Table 10.5** Information exchange and associations between objects of the Protection use case.

| Object name | Description |
| --- | --- |
| Microgrid Measurements and Status (MG Meas. and Status) | Includes voltage, current, frequency, and power (active, reactive) measured at each actor, and the status of the actors, including on/off status, DERs' operation modes (primary DER or as a non-regulating source), as well as other operation status indicators. |
| Microgrid Control Commands (MG Cont. Comm.) | These control commands to microgrid actors (DERs, microgrid EMS, loads, switching devices, protection relays, islanding schemes, and synchronization relays). These commands define each DER's control mode, real and reactive power dispatch, loading for controllable loads, frequency and voltage setting points, islanding, and resynchronization. |
| Protection settings | Settings sent by protection controller to protection relays based on the current operation conditions. |
| Tripping commands | The tripping signal is generated by the protection relay logic and sent to the circuit breaker to trip. |
| Microgrid topology | After circuit tripping, the microgrid topology may change. The new topology information must be sent by to MC and protection. |

Source: Reproduced with the permission of ORNL.

**Table 10.6** Regulations of the Protection use case.

| Regulation | Description |
| --- | --- |
| IEEE Std 1547, series | Standard for interconnection to the area EPS at the PCC. |
| Interconnection agreement | Defines interconnection terms and conditions such as: interconnection studies, operations rules, constructions, safety regulations, maintenance policies, access, boundary limits, disconnect circuit breakers/switching/isolation devices, conflicts in agreements, disconnection, customer generator billing and payment, insurance, customer-generator indemnification, limitation of liability, contract termination, permanents, survival rights, assignment/transfer of ownership of the customer-generator facility, telecommunications, tele-control, and tele-metering. |
| ISO market rules | Defines rules when microgrid participates in markets. |
| Metering regulations | Defines regulations when microgrid exports to AEPS. |

Source: Reproduced with the permission of ORNL.

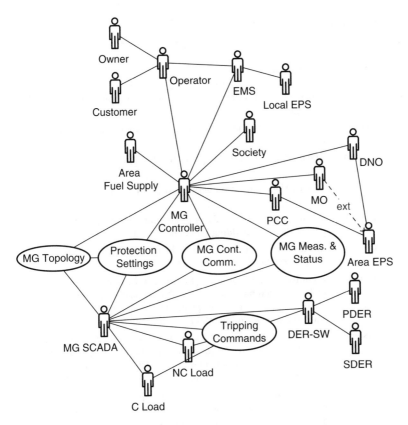

**Figure 10.2** The Protection use case diagram.

## 10.3 Intentional Islanding

This use case defines the function when a microgrid disconnects from the AEPS in a planned way. The microgrid requires permission from the system operator for intentional islanding. The MC, through the EMS and MG SCADA, initiates the process of intentional islanding transition by dispatching islanding transition commands to the actors of the microgrid. The microgrid EMS estimates the microgrid load level and the available generation capacities, shed and/or reduce the lower priority loads, re-dispatch the real and reactive power outputs of each DER and energy storage unit so that there is no import/export of real/reactive power between the microgrid and the AEPS at Point of Common Coupling (PCC). At this stage, the power flow at the PCC is reduced to minimum, and therefore will be minimum impact on both the microgrid and AEPS.

The actor roles within the use case are described in terms of the actor attributes which include name, type and description in Table 10.7. The information exchange model is described in Table 10.8 and the relevant standards are explained in Table 10.9. The use case diagram is shown in Figure 10.3.

**Table 10.7** Attributes of the Intentional Islanding use case.

| Name (diagram label) | Type | Description |
|---|---|---|
| Area Electric Power System (Area EPS) | System | The electrical power system that normally supplies the microgrid through the point of common coupling. |
| Local Electric Power System (Local EPS) | System | The electrical power system on the customer's side of the PCC. |
| Area Fuel Supply (AFS) | System | The fueling system that supplies the microgrid. |
| Point of Common Coupling (PCC) | System | The interface substation between the AEPS and the microgrid. |
| Critical Load (C Load) | Device | The highest priority loads within the microgrid. These loads are not part of the load shedding schemes. |
| Non-Critical Load (NC Load) | Device | The lowest priority loads within the microgrid. These loads may be left unserved in favor of critical loads. |
| Microgrid Controller (MC) | System | A control system that is able to dispatch the microgrid resources including opening/closing circuit breakers, changing control reference points, changing generation levels, and coordinates the sources and loads to maintain system stability. |
| Microgrid SCADA (MG SCADA) | System | Provides the data acquisition and telecommunication required for the microgrid controller functions. It collects real-time data from each microgrid actor, and executes control actions such as economic dispatch commands, circuit breaker controls/status. |
| Primary DER (PDER) | Device | The distributed energy resources participating in voltage regulation. PDERs could be a generator and energy storage system. |
| Switching Device (DER-SW) | Device | The DER-SW can disconnect DER within the microgrid. The DER-SW can receive control signals from the MC and can inform the MC of its status through SCADA. |
| Market Operator (MO) | System | The MO accepts bids from MG, in its AEPS and dispatches MG sources to provide energy and ancillary services. The MO may be part of the AEPS or may be a separate entity. |

Source: Reproduced with the permission of ORNL.

**Table 10.8** Information exchange and associations between objects of the Intentional Islanding use case.

| Object name | Description |
|---|---|
| Microgrid Measurements and Status (MG Meas. and Status) | Includes voltage, current, frequency, and power (active, reactive) measured at each actor, and the status of the actors, including on/off status, DERs' operation modes (primary DER or as a non-regulating source), as well as other operation status indicators. |
| Microgrid Control Commands (MG Cont. Comm.) | These control commands to microgrid actors (DERs, microgrid EMS, loads, switching devices, protection relays, islanding schemes, and synchronization relays). These commands define each DER control mode, real and reactive power dispatch, loading for controllable loads, frequency and voltage setting points, islanding, and resynchronization. |
| Islanding request | A signal sent from MC through the microgrid SCADA to system operator requesting permission for intentional islanding. |
| Islanding request status | System operator allow/not-allow for islanding request. |

Source: Reproduced with the permission of ORNL.

**Table 10.9** Regulations of the Intentional Islanding use case.

| Regulation | Description |
| --- | --- |
| IEEE Std 1547 series | Standard for interconnection to the area EPS at the PCC. |
| Interconnection agreement | Defines interconnection terms and conditions such as: interconnection studies, operations rules, constructions, safety regulations, maintenance policies, access, boundary limits, disconnect circuit breakers/switching/isolation devices, conflicts in agreements, disconnection, customer generator billing and payment, insurance, customer-generator indemnification, limitation of liability, contract termination, permanents, survival rights, assignment/transfer of ownership of the customer-generator facility, telecommunications, tele-control, and tele-metering. |
| ISO market rules | Defines rules when microgrid participates in markets. |
| Metering regulations | Defines regulations when microgrid exports to AEPS. |

Source: Reproduced with the permission of ORNL.

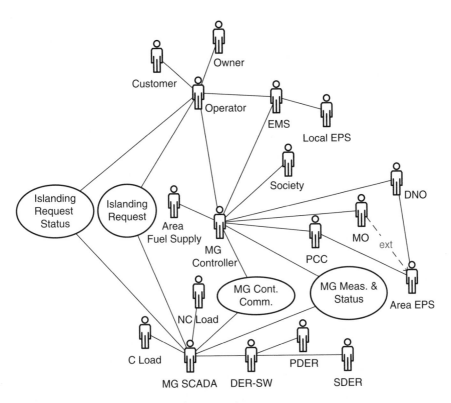

**Figure 10.3** The Intentional Islanding use case diagram.

# 11

# Testing and Case Studies

This chapter discusses testing procedures for the microgrids and their controllers. Some validation results for the modeling, control, monitoring, and protection strategies mentioned in the previous chapters are also presented. The tests were performed on the benchmark models and with field results obtained from the campus microgrid and the utility microgrid in Chapter 2. Four different implementations on a real-live microgrid are discussed in detail. These applications are (i) Energy Management Systems (EMS) economic dispatch, (ii) voltage and reactive power control, (iii) microgrid islanding and (iv) real-time (RT) system using the International Electrotechnical Commission (IEC) 61850 communication protocol.

## 11.1 EMS Economic Dispatch

This section demonstrates the implementation of the design guidelines [41, 60] of the microgrid systems on the campus microgrid's Open Access to Sustainable Intermittent Sources (OASIS) subsystem.

### 11.1.1 Applicable Design on the Campus Microgrid

The OASIS system at the Campus was identified as the microgrid system upon which the EMS and business case frameworks are implemented. The 480 V three-phase system comprises solar Photovoltaic (PV) generation, an Energy Storage System (ESS), Electric Vehicle (EV) loads, and is connected to the utility grid through a local utility (Area Electric Power System, AEPS) substation. The single-line diagram of the OASIS microgrid is shown in Figure 11.1.

The ESS consists of four lithium-ion battery storage groups with a total capacity of 500 kWh. The PV system comprises a large array of 968 PV panels with a maximum of 250 kW generation capacity. The interface from this DC (direct current) bus is a four-quadrant 280 kVA grid-tied and island-able power converter system.

The load primarily consists of four EV charging stations: two level 2 charging stations, and two DC fast-charging stations. The load also comprises the consumed energy to maintain the charging stations, the battery management system, the controller and the Supervisory Control and Data Acquisition (SCADA) in operation.

*Microgrid Planning and Design: A Concise Guide*, First Edition. Hassan Farhangi and Geza Joos.
© 2019 John Wiley & Sons Ltd. Published 2019 by John Wiley & Sons Ltd.

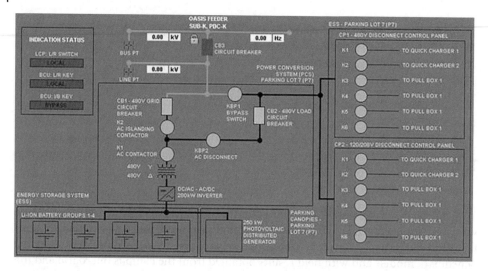

**Figure 11.1** Single line diagram of the campus OASIS microgrid. Source: Ross and Quashie 2016 [69].

The Decentralized Energy Management System (DEMS) acts as the EMS in the OASIS microgrid. The role of the DEMS is to manage the exchange of energy between the microgrid and the Electric Power System (EPS). The objective of the DEMS is not to curtail PV generation if the generation is high, but instead to firm the PV generation if it is low. In other words, it attempts to maintain a minimum PV output as seen by the EPS at the Point of Common Coupling (PCC) if it suddenly drops, and not to reduce the peak power. The microgrid does not obtain a direct financial benefit through arbitrage or other means. It operates on the secondary and tertiary level where it performs a seven-day forecast for the generation from the PV distributed energy resource (DER), and based on this information optimizes the charge/discharge dispatch of the ESS and demand response through the loads.

### 11.1.2 Design Guidelines

The EMS was designed to conform to the following guidelines:

- Capability of simultaneously optimizing multiple conflicting objectives;
- System agnostic generic formulation that is applicable to any system, customizable and adaptable to optimize various objectives;
- Ability to manage the optimal dispatch set-points of the DERs while maintaining power balance without exceeding operating limits;
- Ability to produce results that are self-contained and computationally tractable;
- Allow planning of business cases for microgrid technologies;
- Analyze the investment options for the microgrid and determine the return on investment to the stakeholders; and
- Consolidates all the objectives and guidelines into an encompassing framework.

### 11.1.3 Multi-Objective Optimization – Example

#### 11.1.3.1 System Description

This section describes how the OASIS system parameters and profiles are modeled in the multi-objective optimization (MOO) EMS. Data from the PV generation and total load were used in this study, along with the power at the EPS for comparison with the DEMS. Two profile periods were used in this study:

from November 26, 2015 to January 8, 2016, and
from January 12, 2016 to February 25, 2016.

These two periods were selected for the completeness of the measured parameters (i.e. continuous data were available for all desired parameters without glitches or incorrect measurements). In order to compare the results of the EMS to the existing DEMS, a 15-minute dispatch interval is used.

The residual power in the microgrid is calculated to be:

$$P_{res}(t) = P_{load}(t) - P_{pv}(t) \tag{11.1}$$

where

$$P_{load}(t) = P_{EV1}(t) + P_{EV2}(t) + P_{EV,fast1}(t) + P_{EV,fast2}(t) + P_{sys}(t) \tag{11.2}$$

is the sum of all the EV charging load plus the power to maintain the system.

While there is a high penetration of renewable PV generation, the three controllable resources included in the study are: the ESS, the sheddable load (EV chargers), and the EPS. Although the EPS is not directly controllable as it acts as the isochronous resource to maintain power balance, the formulation is established to consider it as a resource since there is a value associated with the amount of power and energy that is imported/exported.

Both the ESS and the PV system are connected at the same DC bus; therefore, the total amount that the ESS can charge/discharge to the microgrid is dependent on the output of the PV system. Furthermore, the amount of power that it can charge/discharge is dependent on the capacity and state of charge (SoC) of the ESS. Therefore, the power limits of the ESS are:

$$P_{ESS,min} = \max \left\{ -250 - P_{pv}(t), -\frac{E_{ESS} - e_{ESS}(t-1)}{dt} \right\} \tag{11.3}$$

$$P_{ESS,max} = \min \left\{ 250 - P_{pv}(t), \frac{e_{ESS}(t-1)}{dt} \right\} \tag{11.4}$$

The amount of curtailable controllable load is dependent on whether or not EVs are charging, and the amount that they are charging. Therefore, the power limits of this resource are:

$$0 \leq P_{load,shed}(t) \leq P_{EV1}(t) + P_{EV2}(t) + P_{EV,fast1}(t) + P_{EV,fast2}(t) \tag{11.5}$$

The valuation functions for each resource to optimize the identified objectives are described in the following subsection.

### 11.1.3.2 Optimization Formulation

In order for the identified benefits to be congruent with the MOO EMS formulation, they must be expressed in quadratic form. Each objective is addressed by each resource that can help achieve this benefit, and they are quantified and evaluated through identified metrics within the microgrid, utility, and local environment. It should be noted that because the electric power from the EPS is generated through hydro, and the local DER is renewable, greenhouse gas emission is not included as an objective since no further benefits can be achieved through the dispatch.

**EPS**   Three objectives are employed with the EPS resource: cost of energy, energy losses, and power fluctuations. The cost of energy is determined by the total sum of energy consumed by the microgrid over the span of a month. The cost per kilowatt-hour is constant (0.106 $/kWh), unless the total consumption is greater than 14 800 kWh, in which case every kilowatt-hour of energy above this amount is charged at a higher rate (0.1579 $/kWh). The total cost of the energy consumed by the microgrid is shown as the red plot in Figure 11.2, where the jump in price is at the average power that would consume 14 800 kWh in a month.

The energy losses are reduced by the fact that locally produced energy does not need to travel far from the centralized generation stations, thus saving energy through transmission losses. With an estimated loss of 3% through the transmission network, the energy losses based on the power dispatch are:

$$C_{loss}(t) = (0.106 \times 0.03) \cdot p_{EPS}(t)dt \tag{11.6}$$

The total cost of the energy lost by importing power to the microgrid is shown as the $C_{loss}$ plot in Figure 11.2.

The benefits accrued from the power fluctuation are based on the fact that the utility will pay more for firm renewable generation than volatile renewable generation. Therefore, a penalty is associated with any EPS power set-point that deviates from the mean power (as calculated from the past 24-hours), and is equal to the difference in these costs (0.129 − 0.099 = 0.03 $/kWh). The total cost of the power fluctuation deviated from the mean power at the PCC is shown as the $C_{fluc}$ plot in Figure 11.2.

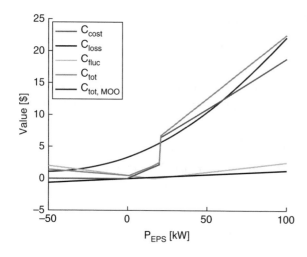

**Figure 11.2** Total valuation functions and quadratic approximation for multi-objective optimization approach. Source: Ross and Quashie 2016 [69].

The scalarization process for MOO problems allows each objective to be summed. Since each of these objectives is commensurable, the total valuation is the summation of each individual objective function. This results in a piecewise linear function (as shown in the $C_{tot}$ plot in Figure 11.2). A curve-fitting approach is used to represent these valuation objectives into quadratic form that is used in the EMS formulation (shown as the $C_{tot,MOO}$ curve in Figure 11.2).

*Load – Reliability*   The value of the reliability is derived from the alternative to not charging EVs. In other words, it is the cost per kilowatt-hour of gas that would have otherwise been consumed through the microgrid. This is calculated to be 0.43 $/kWh, and thus the cost of curtailing the load is a linear function of the amount of load that is shed, as shown in Figure 11.3.

*ESS – Fluctuations*   The ESS can help mitigate any power fluctuations from the PV generation by reacting to its output and either charging or discharging based on its mean power output at the DC bus of the power conversion system (PCS) and the EV charging from the past 24-hours. The calculation of the mean power from the past 24-hours enables the system to adapt to the seasonal effects of PV generation as well as the EV charging pattern. The EPS provides the remainder of the fluctuations, and a quadratic curve of best fit is determined around this mean power ($P_m$ in Figure 11.4).

*ESS – Reliability*   The reliability of the microgrid can be maintained through provisions of the amount of energy stored in the ESS. Through an analysis of the load data, only 167 kWh is required to maintain power to the microgrid loads on average, which is also during a winter season (i.e. low PV output). Since this is only a state of charge (SoC) of 33%, the approach taken here is to make the desired SoC a function of the ratio of load to PV output from the past 24-hours. In this way, during seasons of low PV output, a greater SoC will yield more reliability to power the loads; during seasons of high PV output, the ESS will keep some capacity to charge so that less PV is curtailed or exported to the EPS that may result in high power fluctuations or peaks. The value for reliability

**Figure 11.3** Value (cost) of curtailing electric vehicles when charging within a 15-minute period. Source: Ross and Quashie 2016 [69].

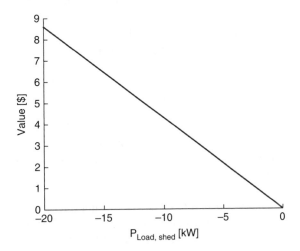

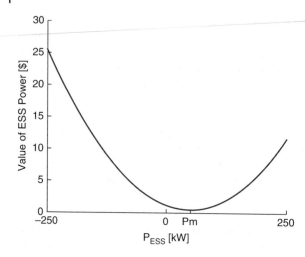

**Figure 11.4** Value of power from ESS to maintain constant power output based on the mean power of the past 24-hours ($P_m$). Source: Ross and Quashie 2016 [69].

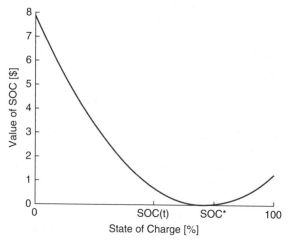

**Figure 11.5** Value of the ESS' state of charge for reliability. Source: Ross and Quashie 2016 [69].

is the same used as for the loads, and the quadratic curve of best fit is again used to determine the penalty for deviating from this desired SoC, as shown in Figure 11.5.

While it is the SoC that determines the reliability, the power dispatch of the ESS can be utilized to either charge or discharge the energy to operate closer to the desired SoC. Figure 11.6 shows the power function of the ESS for the example shown in Figure 11.4. The power is limited by the DER's limits (as described in Section 11.1.3.1), and the power valuation is determined by:

$$C_{Pess,R}(p_{ess}(t)) = |E^*_{ESS} - p_{ESS}(t)/dt| \times 0.43/365 \qquad (11.7)$$

The 0.43 term is the value of reliability as determined by the loads, and the cost is divided by 365 because it is multiplying by the probability of a power outage (assumed to be one day per year overall).

***Summary of Quadratic Parameters*** Table 11.1 shows the overall results for the quadratic parameters as calculated at time for November 26, 2015 as a summation of all quadratic valuation functions. It is important to note that many of the parameters are updated

**Figure 11.6** Value of ESS' power output to reach desired SoC for reliability. Source: Ross and Quashie 2016 [69].

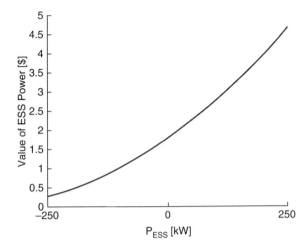

**Table 11.1** Example of quadratic valuation functions of each DER for the MOO EMS framework.

| DER | $\alpha$ ($/kWh²) | $\beta$ ($/kWh) | $\gamma$ ($) | $P_{min}$ (kW) | $P_{max}$ (kW) |
|-----|-------------------|-----------------|--------------|----------------|----------------|
| ESS | 0.000 290 8 | −0.019 79 | 0 | −64.5 | 250 |
| EPS | 0.000 244 1 | −0.000 382 5 | 0.735 9 | −1000 | 1000 |
| Load | 0 | 0.43 | 0 | 0 | 0.899 7 |

Source: Ross and Quashie 2016 [69].

for each dispatch interval, however this table simply shows the results of the quadratic parameters that are congruent with the formulated MOO EMS.

### 11.1.4 Results and Discussion

#### 11.1.4.1 Comparison to Existing Campus DEMS

The two periods of analysis were run through the MOO EMS and are compared to the profiles resulting from the existing DEMS. Figures 11.7 and 11.8 show the respective power import at the microgrid's PCC as well as the mean power import from the past 24-hours ($P_{mean}$). In both cases, one can easily see that not only are the power fluctuations greater with the DEMS profile ($P_{EPS,DEMS}$), but the peak power is significantly less for both importing and exporting power with the MOO EMS ($P_{EPS,MOO}$).

Table 11.2 shows the maximum and minimum power import/export between the microgrid and the EPS. These results show that the peak power is almost reduced by half by utilizing the MOO EMS instead of the DEMS.

Neither EMS shed the load in the dispatch; so the reliability is difficult to analyze directly through the results since there was never an incident with a disconnection from the utility. Therefore, the SoC of the ESS will be analyzed for its provisional ability to maintain an island operation. The SoC profile resulting from the DEMS ($E_{ESS,DEMS}$), the MOO EMS ($P_{ESS,MOO}$) and the desired SoC as determined from section "ESS – Reliability" ($E_{ESS,des}$) are shown in Figures 11.9 and 11.10 for both time analysis periods, respectively.

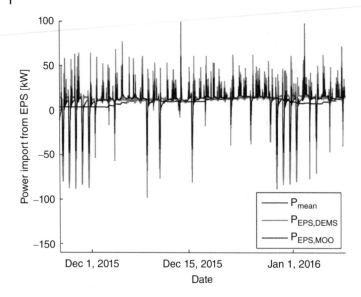

**Figure 11.7** Comparison of power import from the EPS for both EMSs for the first time period analysis. Source: Ross and Quashie 2016 [69].

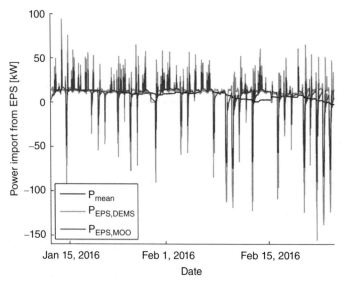

**Figure 11.8** Comparison of power import from the EPS for both EMSs for the second time period analysis. Source: Ross and Quashie 2016 [69].

In terms of reserve provision, the DEMS typically maintains a higher SoC than the MOO EMS. However, the calculated desired SoC shows that this is not necessary to maintain an islanded scenario. The strategy of the DEMS to maintain a high SoC may not be favorable in the summer months with a large amount of PV generation since the excess generation would be exported to the EPS, thus potentially resulting in higher power peaks and fluctuations.

**Table 11.2** Peak power import/export at the microgrid's PCC for both EMSs.

| EMS | $P_{eps,min}$ (kW) | $P_{eps,max}$ (kW) |
|---|---|---|
| DEMS, T1 | −98.3 | 101.7 |
| MOO, T1 | −52.0 | 61.7 |
| DEMS, T1 | −155.7 | 94.4 |
| DEMS, T2 | −89.5 | 59.4 |

Source: Ross and Quashie 2016 [69].

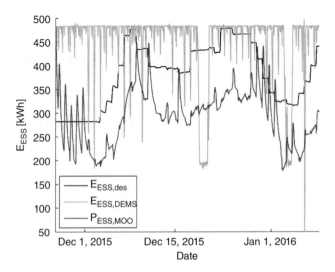

**Figure 11.9** State of charge for the ESS for both EMSs for the first time period analysis. Source: Ross and Quashie 2016 [69].

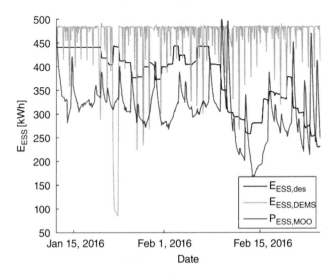

**Figure 11.10** State of charge for the ESS for both EMSs for the second time period analysis. Source: Ross and Quashie 2016 [69].

By contrast, the MOO EMS utilizes the ESS more, as is reflected by the amount of variation in the SoC. The SoC does not directly copy the desired SoC profile, this is due to the trade-off in other objectives during the optimization process. That is, if the objective valuation functions put a relatively greater weight on the reliability benefit, then the SoC would track the desired SoC much closer. However, as discussed previously, the system is still able to reduce the peak power and fluctuations while maintaining sufficient reserve in case of an islanding event (only in a single time period did the reserve go below the identified minimum power: 164.5 kWh).

### 11.1.4.2 Business Case Overview

The EMS was subjected to a cost-benefit analysis and the results are detailed in Table 11.3. Key financial metrics such as the Present Value Ratio (PVR) and Internal Rate of Return (IRR) are used in determining the financial viability (profitability) of an investment in the EMS. PVR is the ratio of the Net Present Value (NPV) of the microgrid benefit/revenue to the present value of the investment cost. A PVR greater than one indicates the profitability of an investment, which is unprofitable when its PVR less than one. Herein, the economic merits of the MOO EMS are compared with that of the existing system based on the outputs of both systems. The microgrid benefits discussed earlier were considered in the evaluation. The estimated impacts/benefits were assigned to their corresponding stakeholders (microgrid owner/customer, distribution network operator and the society) as shown in Table 11.3 with the base unit being the total cost of the existing system.

For the purposes of the business case, PV measurement (forecast) over summer periods were super imposed on measured data to present a fair assessment of the system over an entire year. In both cases, the OASIS system happens to be a net generator due to the high capacity of the PV and ESS installed relative to its load. This

**Table 11.3** Comparison of business cases for the two EMSs.

| Stakeholder/Actor | | Microgrid controller NSMG (p.u.) | Existing microgrid controller (p.u.) | |
|---|---|---|---|---|
| MG owner/customer | Power fluctuation (firm energy) | 0.148 79 | 0.225 03 | |
| | Energy cost | −0.005 15 | −0.005 91 | |
| | Investment cost | 0.120 48 | 0.779 79 | |
| | Total | 0.264 12 | 1 | |
| MG owner PVR | | | | 6.098 92 |
| Savings MG owner/customer (%) | | | | 73.56 |
| Distribution network operator (DNO) | Investment deferral | −0.010 60 | −0.010 90 | |
| | Improved efficiency | 0.001 07 | 0.001 09 | |
| | Total | −0.004 81 | −0.004 94 | |
| Total savings DNO | | | | −2.771 03 |
| Society | Emission EV | −0.007 67 | −0.007 67 | 0 |

Source: Ross and Quashie 2016 [69].

results in a negative cost of energy (revenue) as indicated in Table 11.3. Also more energy is exported by the existing system to the grid with higher fluctuation when compare to the proposed system as observed in Figures 11.7 and 11.8. Thus, the existing system though it generates more revenue, attracts higher penalty for deviating from its firm energy. Both systems do not share any load in the event of islanding hence no cost was incurred for non-delivered energy. Implementation of the proposed system yields a significant saving to the microgrid owner. This is primarily due to the relatively low cost of the system which provides less functionality than the existing system but enough for a microgrid application of this scale. In the same perspective, the profitability index of the proposed system is relatively high and justifies its investment.

Also, the utility or distribution network operator seems to accrue more benefit in relation to system efficiency improvement in the analyzed case. With respect to investment deferral, the existing system provides more benefit as shown in Table 11.3. Furthermore, the total benefit provided by the existing system to the distribution network operator is more than that provided by the MOO EMS. However, the magnitude of these benefits in comparison to the accumulated benefit to the microgrid owner is relatively low. Based on the assumption made in determining savings from emission reduction and the fact that both systems provide the same amount of energy to the EVs, the society accrued the same benefit in both cases.

## 11.2 Voltage and Reactive Power Control

### 11.2.1 VVO/CVR Architecture

Sample values from smart meters (SMs) could be used to construct real-time load profiles for Volt-VAR optimization (VVO)/conservation voltage reduction (CVR) engines (Figure 11.11). SMs in new distribution networks provide accurate and real-time measurements of service quality levels at termination points.

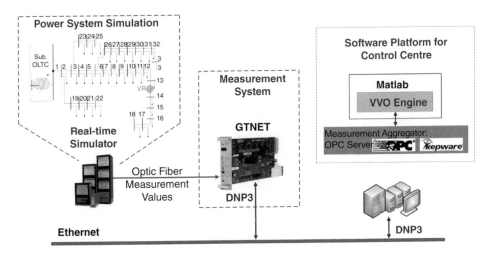

**Figure 11.11** Volt-VAR optimization Engine (VVOE) platform. Source: Manbachi et al. 2015 [22]. Reproduced with the permission of the IEEE.

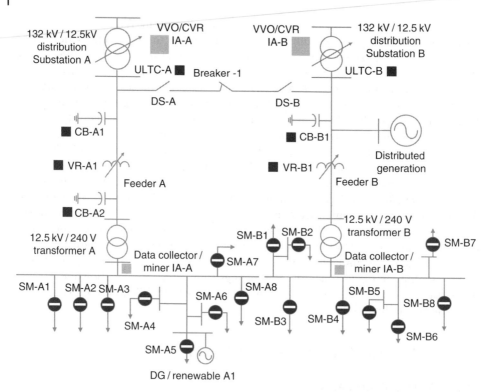

**Figure 11.12** IA-based VVO/CVR primary structure in a distribution network (IA, intelligent agent; OLTC, on-load tap changer; DS, disconnector; CB, capacitor bank; VR, voltage regulator; SM, smart meter). Source: Manbachi et al. 2014 [18]. Reproduced with the permission of the IEEE.

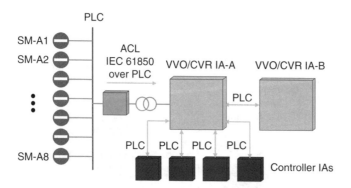

**Figure 11.13** Communication structure of proposed VVO/CVR system, using PLC, agent language programming, access control list (ACL), and IEC 61850 standard. Source: Manbachi et al. 2014 [18]. Reproduced with the permission of the IEEE.

To capture, analyze, and exchange data points related to feeder status, a system of intelligent agents (IAs) as well as a communication network is required. The approach constructs real-time load profiles that are fed to integrated VVO/CVR engines through an agent-based distributed command and control system. The communication platform uses IEC 61850 and narrowband power line communication (NB-PLC) technology. IAs are tasked with applying appropriate configurations for relevant assets within a microgrid or utility's distribution network.

Figure 11.12 depicts the primary structure of the proposed real-time IA-based VVO/CVR system. Two distribution feeders are paralleled with a tie-breaker (Breaker-1) to supply residential customers. Both feeders have Volt-VAR control devices at the medium voltage (MV) side of their 12.5 kV/240 V transformers. Each feeder has residential loads and each residential termination point has an SM which captures load data (and if need be, other sample value data). SMs send data to other network nodes from the low voltage (LV) side to the MV side.

Here, the distributed command and control scheme can be visualized as three IAs controlling integrated VVO/CVR application. This classification is fully dependent on agent functionalities, tasks, and the location of the component within the distribution network. Thus, the required data for VVO/CVR are collected, analyzed, and processed by IAs implanted in the system. Such agents are tasked with processing smart metering data, determine probable events that have produced that data, and communicate their findings with the main VVO/CVR IA, which in turn can determine the new configuration settings for VVO/CVR components in the system. The ability to communicate the event, rather than the raw Advanced Metering Infrastructure (AMI) data, will substantially reduce the bandwidth requirement of the communication system, and improves its performance. Figure 11.13 shows the communication structure for VVO/CVR applications using NB-PLC and the IEC 61850 standard.

For instance, assume Volt-VAR optimization Engine (VVOE) sends a command to a voltage and/or reactive power (VAR) injector circuit breaker (CB)-A1 agent (CB in feeder A) to inject 50 kVAR into the system in a specified real-time interval. CB-A1 is an agent who keeps records of its CBs. Hence, it can decide which bank units have to be ON/OFF to cover the 50 kVAR. This can be done based on CB positions and data. If the requested task could not be implemented, CB-A1 will inform the VVOE. The latter would then be responsible for finding an alternative solution for the network using an algorithm shown as a flowchart in Figure 11.14.

## 11.3 Microgrid Anti-Islanding

An inverter's anti-islanding capabilities were tested on a real test distribution feeder test-line [70] configured for microgrid studies. The tests that were conducted are supplemental to current standards of anti-islanding testing, as various experiments were performed that are not all included in Institute of Electrical and Electronics Engineers (IEEE) Standard 1547.1. These tests provide interesting insights to the operation of the system with the DER, as well as the inverter control system. To test and verify the ability of a DER to detect an islanding condition and disconnect itself within a specified time, most follow IEEE 1547.1 *Standard Conformance Test Procedures for Equipment Interconnecting Distributed Resources with EPSs*. The tests performed in this project will supplement standard test procedures by subjecting the system to potential yet realistic situations. The testing results of an inverter system that will be used to interface a solar concentrator DER to the grid are presented. The inverter was tested on a real

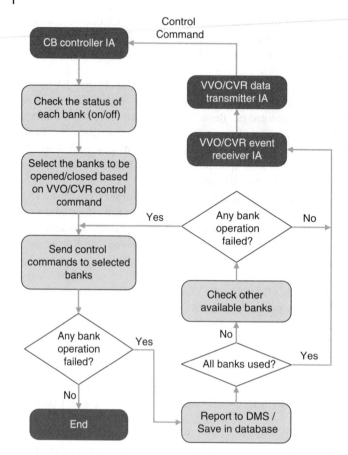

**Figure 11.14** Initial flowchart of CB controller IA operating tasks. Source: Manbachi et al. 2014 [18]. Reproduced with the permission of the IEEE.

test distribution system to validate its anti-islanding detection and to characterize its reaction to such real-system events. Tests that were carried out included: islanding with zero power mismatch between DER and load while varying P (also the quality factor), capacitor switching, motor start-up, and islanding with a motor.

## 11.3.1 Test System

### 11.3.1.1 Distribution System

The inverter was installed on a real distribution test feeder, as shown in Figure 11.15. This test line is a dedicated 25 kV overhead line that is fed from a 120 kV/25 kV, 28 MVA Y-Δ substation transformer, grounded using a zig-zag transformer. The three-phase system is a four-wire solidly grounded system using 477 AL overhead lines with a length of approximately 300 m from the substation. The feeder consists of a CB, series reactance ($X_S$), three single-phase voltage regulators (VRs), and switchable shunt capacitor banks, as denoted in Figure 11.15. A step-down transformer connects the generator (Distributed Generation, DG) the loads (R, $X_L$, $X_C$), and an induction motor (M) to the distribution system at the same PCC.

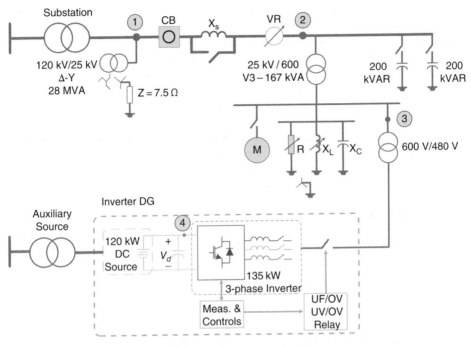

**Figure 11.15** Single-line diagram of the system topology used in the anti-islanding tests. Blue circled numbers indicate measurement points. Source: Ross et al. 2012 [70]. Reproduced with the permission of the IEEE.

The series reactance $X_S$ comprises two 100 kVA, 14.4 kV/347 V series connected distribution transformers to emulate the voltage drop associated with a long rural distribution line. The loads R, $X_L$, and $X_C$ are connected in wye and are independently controllable, which will vary for the tests to consume the required power and reactive power, thus also changing the quality factor of the load. Although the voltage regulators VR are active in these studies, they simply act as autotransformers with a fixed tap setting since the controls act much slower than the timeframe of the islanding and tripping. They simply ensure a proper voltage to the feeder before the system is islanded. Measurements are made and recorded on a TEAC LX-120 data logger and signal analyzer. Voltage measurements were made at Point 1, and voltage and current measurements were made at Points 2, 3, and 4 as shown in Figure 11.15.

Simulations always have assumptions in models. Models used in simulations cannot encompass all phenomena of a real system since: (i) models make simplifications based on the purpose of the simulation (timescale) to reduce computation time and complexity; (ii) simulation solvers operate on the discrete timeframe, and often use reduced order solvers; and (iii) new theories and models are consistently being developed and presented that better represent a system. In addition, from the point of view of inverters, the controllers are essentially "black boxes," such that their internal (often proprietary) controls are difficult to model.

Emulations, which imitate systems with a representative physical system, are other means to perform tests for validation. The advantage of physical emulations is that many phenomena associated with the real devices are observed. The disadvantage is that typical laboratory bench tests may not have the same power ratings, or if they do, they can only emulate part of the power system (i.e. it will be difficult to emulate true transmission lines or all power equipment in a lab setting).

The main advantage of testing on a real distribution test system over other methods is that all phenomena associated with a distribution system are encompassed, and they provide realistic values of an actual system without any simplifications or assumptions. This can give rise to interesting results that may not be perceived in simulation or emulation. Actual distribution equipment is used, and the tester has control over the topology, loads, generation, and any device that one wishes to include in the tests. Of course, this can be achieved by testing on a real system, however a test distribution system provides more flexibility in the variety of tests one can perform, since some tests could affect utility customers and potentially damage personal equipment.

### 11.3.1.2 Inverter System

The inverter was connected to a DC source that was obtained from a different feeder than the distribution test system. That way, the inverter was interfacing a separate power source that is isolated from the distribution test feeder, thus representing an independent DER. The power source assumed constant operating conditions in steady state (constant power at constant voltage) before the testing events occur.

The inverter was a pre-assembled package that contained all necessary controllers, harmonic filters, and components. Typical utility requirements for decentralized generators are to protect against three-phase under-voltage (UV) and over-voltage (OV) as well as under-frequency (UF) and over-frequency (OF) for islanding of the distribution system. The internal relay settings for UV/OV and UF/OF are provided in Table 11.4.

### 11.3.2 Tests Performed and Results

The following subsections describe the various tests that were performed that supplement the current standards of testing for anti-islanding functionality along with their intended purposes and the results obtained.

**Table 11.4** Relay U/O voltage and U/O frequency threshold settings.

| Parameter | Threshold settings |
| --- | --- |
| Under voltage | 0.75 p.u. for 1 s |
|  | 0.9 p.u. for 2 s |
| Over voltage | 1.2 p.u. for 0.1 s |
|  | 1.06 p.u. for 120 s |
| Over frequency | 63.5 Hz for 0.01 s |
|  | 63 Hz for 5 s |
| Under frequency | 55.5 Hz for 0.01 s |
|  | 56.5 Hz for 0.35 s |
|  | 57 Hz for 2 s |

Source: Ross et al. 2012 [70]. Reproduced with the permission of the IEEE.

### 11.3.2.1 Nuisance Tripping

The purpose of these tests is to analyze how the system reacts to OV and UV transients. Such transients can occur by switching capacitor banks or connecting/disconnecting a large amount of load or generation in a very short period of time. Again, a small power mismatch is used in these tests, as a worst-case scenario. Although a short response time to an islanding event is favorable, too short of a response time can lead to nuisance tripping during system transients. Two sets of tests were performed to validate the functionality of the inverter during non-islanding events on the grid: capacitor switching and motor starting.

*Capacitor Switching*  The 200 kVAR capacitor bank will be switched on the system to increase the voltage (OV transient), and be disconnected to decrease the system voltage (UV transient) after the system has stabilized to a new steady state condition. The first set of tests bypasses the series reactance $X_S$, but the second set of tests includes them to represent a long distribution line between the substation and loads/DER. The resonance created with this impedance and the capacitor bank is observed. The inverter output power is initially 105 kW when the 200 kVAR capacitor bank is switched, then it is increased to 115 kW when the 400 kVAR capacitor bank is switched. For these tests, the induction motor M is disconnected.

Nuisance tripping was found to occur when the series impedance $X_S$ was not connected to the line, and the 200 kVAR capacitor bank was switched on. This caused the current from the 25 kV distribution substation to reach as high as 80 $A_{RMS}$, and finally settle after 37 ms of the capacitor bank being connected as shown in Figure 11.16. The inverter must have recognized this from the high current it was drawing (the voltage and frequency remained within its proper operating regions), and disconnected the DER within 6.2 ms.

When the series impedance is connected to the system, the inverter still trips when the capacitor bank is connected. However, the inrush current from the substation is lower (17 p.u. as opposed to 19 p.u., on a 100 kW power base). The added impedance on the system also creates more harmonics and current distortion on the system, until the voltage and currents stabilize as shown in Figure 11.17. During the test when the capacitor

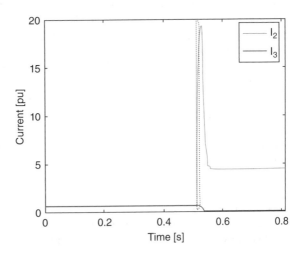

**Figure 11.16** Per unit currents from the substation (blue) and DER (red) during capacitor bank connection, on a base power of 100 kW. Source: Ross et al. 2012 [70]. Reproduced with the permission of the IEEE.

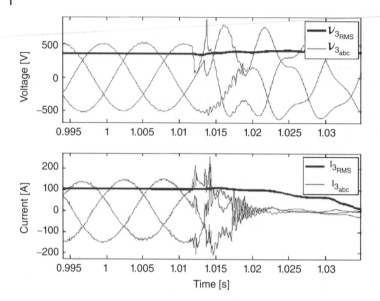

**Figure 11.17** Voltage and current waveform (blue) and RMS (red) values from the inverter during capacitor bank connection (with a series impedance $X_s$).Source: Ross et al. 2012 [70]. Reproduced with the permission of the IEEE.

bank is disconnected, the voltage dips down, but both the voltage and frequency remain within proper operating regions and no excess current is drawn from the inverter. Therefore, the inverter does not disconnect.

*Motor Starting* Current standards only test for passive, linear resistive-inductive-capacitive (RLC) loads. This, however may not be the case in real life as motors, for example, constitute a substantial portion of load on the system. These loads can have complex dynamic behaviors that can affect an islanding protection scheme's ability to appropriately recognize such situations.

This test determines how the inverter reacts to a motor connecting directly to the system in a worst-case scenario; the 200 HP induction motor is directly connected to the 600 V bus without a soft-start. The 200 kVAR capacitor bank and the series inductance are connected in these tests. The motor should draw a large start-up current that may affect the inverter's anti-islanding detection scheme.

During start-up, the voltage of the system sharply decreases as the motor draws significant inrush current. However, this inrush current comes mostly from the substation (rising from 0.6 to 3.8 p.u.), as opposed to the DER, which remains relatively constant at 0.6 p.u., as shown in Figure 11.18. As the motor settles into steady state, the voltage rises back to 1 p.u., and the current from the subsystem settles into a stable operating point. Even in the worst-case scenario for motor start-up (directly connected with no soft-start and full load), the system's voltage and frequency remain within their acceptable operating region, and the inverter does not disconnect.

### 11.3.2.2 Islanding
Three test cases are performed in this study: (i) Varying the DER power with respect to the load power; (ii) Varying the power output of the DER and load, while keeping the

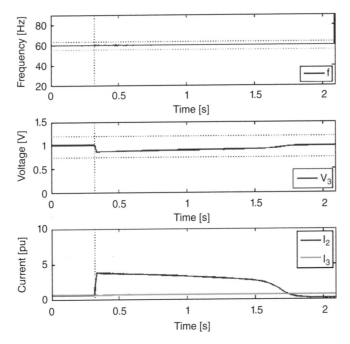

**Figure 11.18** Frequency, voltage and current RMS values during motor start-up (blue for substation current, red for inverter current). Source: Ross et al. 2012 [70]. Reproduced with the permission of the IEEE.

mismatch minimized (close to 0 W); and (iii) Islanding the system with an induction machine. In all cases, the total reactive power of the system was adjusted such that the power factor remained close to unity.

*Varying ΔP*  The first test set analyzes how the inverter's controls react to the islanding condition when the distribution system is overall importing power ($\Delta P < 0$), and exporting power ($\Delta P > 0$), as shown in Test numbers 1–8 in Table 11.5. These tests are performed to verify the functionality of the inverter under typical testing conditions.

The results for the islanding scenarios are shown in Table 11.6, where test numbers are as indicated in Table 11.5. For most of the tests, the voltage and frequency either consistently increased or decreased to outside permissible operating conditions, thus tripping the inverter. For instance, Figure 11.19 shows the voltage and frequency of test number 2 (note that frequencies after islanding are only representative of resonance), and Figure 11.20 shows the voltage and current waveforms. The parameters $t_{UV}$, $t_{OV}$, $t_{UF}$, and $t_{OF}$ refer to the times that the measured voltage or frequency go beyond their limits, as outlined in Table 11.6. The parameter $t_{disconnect}$ refers to the time when the inverter system disconnects itself from the EPS.

For most test cases with a relatively large power mismatch (approximately 25 kW and greater) when importing power, both the voltage and frequency go out of range with UF and UV detection. When the mismatch is zero or when exporting power (such as the case with test numbers 6–8), the voltage remains within the acceptable range before islanding occurs. This is a feature of the positive feedback in the inverter. In a parallel configuration, the voltage is related to the power and the frequency is related to the

**Table 11.5** Power settings for test condition.

| Test number | $P_{DG}$ (kW) | $P_{load}$ (kW) | $\Delta P$ (kW) | $Q_f$ |
|---|---|---|---|---|
| 1 | 100 | 300 | −200 | 0.19 |
| 2 | 100 | 170 | −70 | 0.34 |
| 3 | 100 | 145 | −45 | 0.39 |
| 4 | 100 | 125 | −25 | 0.46 |
| 5 | 25 | 50 | −25 | 1.14 |
| 6 | 100 | 120 | −20 | 0.48 |
| 7 | 25 | 0 | 25 | — |
| 8 | 100 | 70 | 30 | 0.81 |
| 9 | 25 | 25 | 0 | 2.28 |
| 10 | 95 | 95 | 0 | 0.6 |
| 11 | 100 | 100 | 0 | 0.57 |

Source: Rose et al. 2012 [70]. Reproduced with the permission of the IEEE.

**Table 11.6** Islanding time and reason for island.

| Test number | $\Delta P$ (kW) | $Q_f$ | $t_{UV}$ (ms) | $t_{OV}$ (ms) | $t_{UF}$ (ms) | $t_{OF}$ (ms) | $t_{disconnect}$ (ms) |
|---|---|---|---|---|---|---|---|
| 1 | −200 | 0.19 | 10.4 | — | 16.4 | — | 19.2 |
| 2 | −70 | 0.34 | 15.7 | — | — | 10.9 | 39.3 |
| 3 | −45 | 0.39 | 90.9 | — | 5.8 | — | 109.8 |
| 4 | −25 | 0.46 | 80.2 | — | 18.8 | — | 111.3 |
| 5 | −25 | 1.14 | 16.6 | — | 13.7 | — | 115.5 |
| 6 | −20 | 0.48 | — | — | 3.0 | — | 48.5 |
| 7 | 25 | — | — | 10.3 | — | 4.9 | 15.0 |
| 8 | 30 | 0.81 | — | — | 98.8 | — | 134.0 |
| 9 | 0 | 2.28 | — | — | — | 9.8 | 93.8 |
| 10 | 0 | 0.6 | — | — | 59.4 | — | 115.4 |
| 11 | 0 | 0.57 | — | — | — | 21.5 | 47.7 |

Source: Rose et al. 2012 [70]. Reproduced with the permission of the IEEE.

reactive power for the positive feedback as:

$$P = \frac{V^2}{R}Q = V^2\left(\omega C - \frac{1}{\omega L}\right)$$

Therefore, as the power mismatch approaches zero, it is consistent that the positive feedback for the power does not drive the voltage to either extreme.

In terms of determining a relationship between either $\Delta P$ and $t_{disconnect}$ or $Q_f$ and $t_{disconnect}$, it is very difficult to do so using results from real tests for many reasons. First, in these tests, it is difficult to independently vary one parameter while keeping other

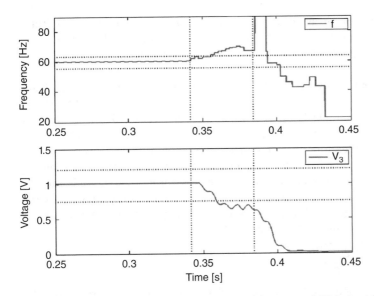

Figure 11.19 Frequency and RMS voltage values of the inverter PCC during islanding (test number 2).Source: Ross et al. 2012 [70]. Reproduced with the permission of the IEEE.

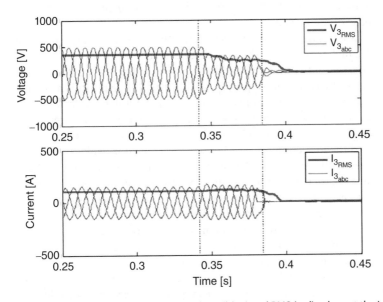

Figure 11.20 Voltage and current waveform (blue) and RMS (red) values at the inverter's PCC during islanding (test number 2). Source: Ross et al. 2012 [70]. Reproduced with the permission of the IEEE.

parameters constant. Since loads must increase in 5 kW (as dictated by the system setup), it is almost impossible to get exact power and quality factors to what one desires. Therefore, plotting either $t_{disconnect}$ vs. $\Delta P$ or $t_{disconnect}$ vs. $Q_f$ would not yield conclusive results – there are too many variables that change from one test to the next, not just $Q_f$ and $\Delta P$. In addition, one is unable to precisely control when the islanding event occurs (for example, one cannot always disconnect when Phase A voltage is at 0°). These variabilities in the tests make it difficult to conclude relationships and trends between certain parameters. This is the drawback of testing on a real system.

*Varying Power Output and Quality Factor*   The second test varies the power of the DER and load while keeping them as equal as possible, for a high load and a low load (as shown in Test numbers 9–11 in Table 11.5). The worst-case scenario for detecting when an islanding event has occurred is when there is virtually zero power mismatch between the DER and the load, therefore operating in a potential Non-Detection Zone (NDZ). An NDZ is a power mismatch between load and generator that causes the anti-islanding detection to be unable to detect and disconnect the DER within the required time.

It is also interesting to note that, while the system's reactive power stays the same, changing the P will change the Quality Factor ($Q_f$) of the system. The quality factor of the load is varied in the studies, which is defined as:

$$Q_f = R\sqrt{\frac{C}{L}} = \frac{\sqrt{Q_L \times Q_C}}{P}$$

where, $Q_f$ is the quality factor of the resonant load; R is the effective load resistance ($\Omega$); C is the effective load capacitance (F); L is the effective load inductance (H); $Q_L$ is the reactive power per phase consumed by the inductive load component (VAR); $Q_C$ is the reactive power per phase consumed by the capacitive load component (VAR); and P is the real power per phase of the resistive

The quality factor is a measure of how under-damped the resonant load is after the system is islanded. The higher the value, the resonance created by the load will be less damped. In other words, the higher the $Q_f$ value, the stronger the tendency of the system to move toward or stay at the resonant frequency. This will affect the positive feedback frequency algorithm to make the inverter operate outside the UF/OF boundaries, if the resonant frequency is close to the system frequency of 60 Hz. Since L and C were constant in all tests, varying P would also vary $Q_f$. For these tests, the series reactance $X_S$ is included in the line, and the induction motor M and shunt capacitor banks are disconnected. The reactive load on the system is adjusted to maintain a power factor of 1.0 p.u.

As is shown in Table 11.6, there does not appear to be any correlation between detection time and either $\Delta P$ or $Q_f$ between test cases 9 and 10. Simply observing either test cases 9 and 11 or test cases 10 and 11, the detection time decreases as the inverter power increases. This is consistent with the findings, which concluded that the worst case scenario for islanding detection is when the inverter approaches full power. The quality factor may not have had too much of an effect on the system because, even though the resonant frequency was approximately 60 Hz (since the power factor was very close to unity), the quality factor was still low, and resonance may not have a significant effect on the system and the islanding detection scheme. In terms of islanding detection, if the system's resonant frequency is close to 60 Hz (LC $\approx [2\pi \times 60]^2$) and its quality factor is high, islanding detection schemes based on frequency measurement may have a larger NDZ.

*Islanding with a Motor*   Detecting the islanding condition with a 200 HP induction motor may prove to be a challenge, since the motor has inertia that, when the system is islanded, helps maintain voltage and frequency (it acts as a generator, supplying energy from its spinning mass).

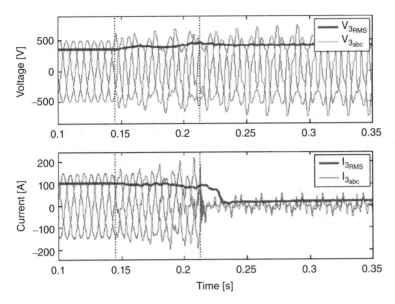

**Figure 11.21** Three-phase voltage and current waveform (blue) and RMS (red) values at the DER PCC during islanded condition with induction machine connected. Source: Ross et al. 2012 [70].

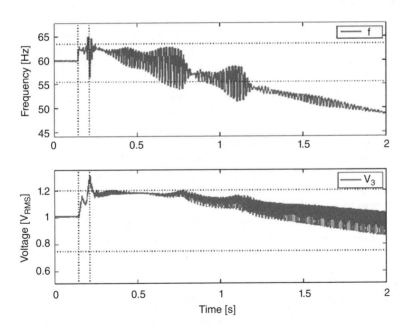

**Figure 11.22** DER PCC frequency and RMS voltage during islanded condition with induction machine connected. Source: Ross et al. 2012 [70].

A worst-case scenario is performed to see how the inverter would react. The load on the induction motor is removed just as the system is islanded. Therefore, the motor would act as a generator just before islanding, helping to maintain system frequency and voltage. For the motor tests, the series reactance $X_S$ is included, and the 200 kVAR shunt capacitor bank is connected.

Figures 11.21 and 11.22 show the voltage and current waveforms, and the rot mean square (RMS) voltage and frequency (respectively) measured from the DER's PCC during system islanding with a motor. The DER was disconnected since it measured an OV. Figure 11.21 show the three-phase voltage and current waveform and RMS values at the DER PCC during islanded condition with induction machine connected. Left and right vertical lines denote when the system is islanded, and when the DER is disconnected, respectively. In Figure 11.22 the DER PCC frequency and RMS voltage are shown during islanded condition with induction machine connected. Horizontal lines denote OV/UV and OF/UF limits, respectively, and the left and right vertical lines denote when the system is islanded.

The voltage first goes over 1.2 p.u. at 15.7 ms after the system is islanded, and it stays within operating frequency range while islanded. The DER is disconnected at 6.82 ms after the islanding event, which means that it remained at a high voltage for approximately 50 ms (much longer than shown in Table 11.5).

## 11.4 Real-Time Testing

Standardization efforts were underway at the time of writing this text for testing the microgrid control system [70]. There are several ways a microgrid control system can be tested depending on the stage of microgrid development as shown in Figure 11.23. The figure shows test approaches for microgrid controllers but is general for any type of asset controller and/or its protection. The first approach that could be taken at the early development stage is that of simulating the grid, the assets, and the microgrid control system using an offline simulation software. This approach provides full coverage of tests (as long as the test conditions can be modeled). Since this approach does not completely capture the microgrid controller (or other asset controllers), another approach called the controller hardware in the loop (CHIL) testing can be used. In this approach actual controllers of the assets and the microgrid can be connected to a simulated electrical network that runs in real time. This approach also allows full coverage for testing; however, real-time implementation of the network will require reduced-order modeling or simplification of some complex models to ensure real time simulation. Power processing, either full-scale or scaled down, can be added to the CHIL test approach. This approach is conventionally called power hardware in the loop (PHIL). This approach provides limited test coverage; however, the actual power conversion aspects of the system can be tested using this approach. CHIL and PHIL approaches are used in the controller selection or pre-commissioning phase of the development. A full rated actual hardware test is the approach that could be taken, however, it has limited test coverage and is expensive to set up.

This section presents two examples of setting up a real-time test bed for microgrid testing, validation, and experimentation. The first example is of integration of real time simulators (RTS) and control hardware to provide a hardware-in-the-loop (HIL) testing platform for commercial equipment such as relays and CB elements. The other example

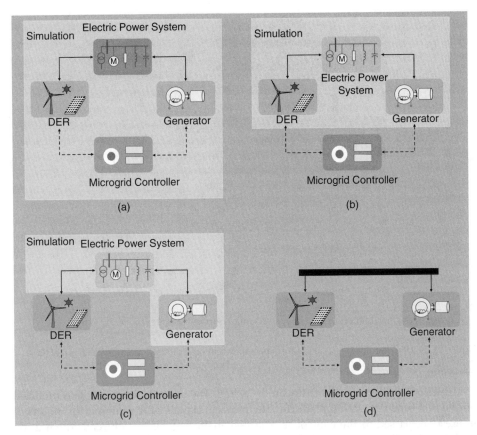

**Figure 11.23** Options for laboratory evaluations of microgrid controller compliance with site-specific requirements: (a) pure simulation, (b) CHIL, (c) CHIL and PHIL, and (d) hardware only. Source: Maitra et al. 2017 [2]. Reproduced with the permission of the IEEE.

also models the microgrid on an RTS, however, each element is modeled as an agent that communicates to the other agent via the IEC 61850 messaging protocol. Both examples are explained next.

### 11.4.1 Hardware-In-The-Loop Real Time Test Bench

The HIL RT test bench was designed to the have the following features:

1. Ability to verify control strategies and corresponding algorithms for the microgrid by digitally implementing it in an industrial hardware platform;
2. Capable of addressing practical issues that do not manifest themselves in offline simulation studies; these practical issues include noise, sampling rate of the digital controllers, sample and hold of sinusoidal pulse width modulation (SPWM) signals, and delay and nonzero bias of analog inputs and outputs.

The overall block diagram of the test bench is shown in Figure 11.24. It involves an industrial control and protection platform that is programmed with the control

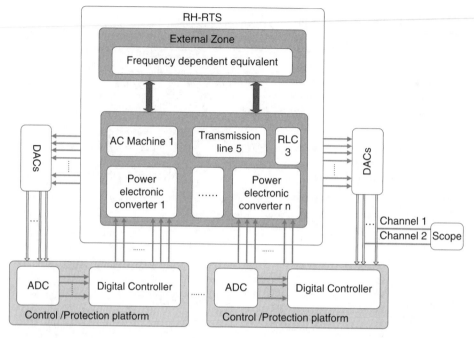

**Figure 11.24** Hardware-in-the-loop platform based on real-time digital simulator and the real time controller. Source: Etemadi et al. 2012 [13]. Reproduced with the permission of the IEEE.

strategies and protection algorithms to be tested. The microgrid as a whole is modeled on the real time simulation platform. The physical signals of the microgrids' points of interest are passed to the controller as scaled analog signals, which reads them using analog to digital converters (ADCs).

Figure 11.25 shows a schematic representation of the HIL environment for the closed-loop performance evaluation of the microgrid of Figure 3.9. The system in Figure 11.25 is composed of:

1. A rack that simulates the power circuitry of Figure 3.9, including the interface voltage sourced converter (VSCs) of the DER units;
2. Input/output (I/O) interface for signal transfer between the simulator and control platform of the DER units;
3. Three real-time (RT) control platforms where each real-time controller (RTC) unit embeds the control algorithms of one DER unit;
4. A function generator that once every second sends a synchronization pulse to each RTC to synchronize the independently generated 60 Hz angle waveforms of the DER units.

Each RTC features a real-time processor communicating with a field-programmable gate-array (FPGA) chip via the backplane. The FPGA chips are used to generate the SPWM signals with the switching frequency of 6 kHz. The proposed controller is discretized with a sampling frequency of 6 kHz and implemented in the real-time processors using C-programming language. The 60 Hz phase angle required for *dq*

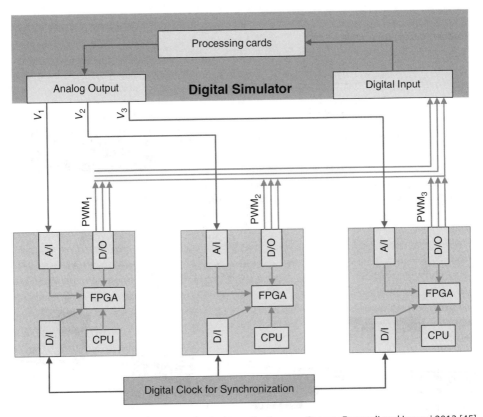

**Figure 11.25** Real-time simulation test bed schematic diagram. Source: Etemadi and Iravani 2013 [45]. Reproduced with the permission of the IEEE.

transformation is generated in each RTC separately and synchronized using the external digital signal generator that emulates a Global Positioning System (GPS). A computer emulates the Power Management System (PMS) by performing power flow analysis and communicating the reference set points to RTCs via Ethernet.

## 11.4.2 Real-Time System Using IEC 61850 Communication Protocol

The HIL microgrid controller test bed with an RTS is shown in Figure 11.26. The test bed models each microgrid element as an agent that communicates to the other agents via the IEC 61850 messaging protocol, in contrast to the physical analog signals in the prior test bed. The IEC 61850 standard, originally put together to ensure interoperability among Intelligent Electronic Devices (IEDs) from multiple manufacturers and simplify commissioning, has gained recent interest in distribution automation due to its real-time, low latency (max 4 ms) generic object oriented substation event (GOOSE) protocol. The standard allows for communication between devices, where a peer-to-peer model for generic substation events (GSEs) services is used for fast and reliable communication between IEDs. One of the messages associated with the GSE services is GOOSE, which allows for the broadcast of multicast messages across

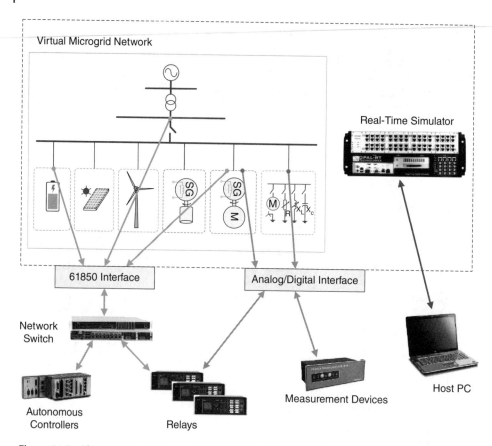

**Figure 11.26** The real-time platform with IEC 61850 and HIL capability.

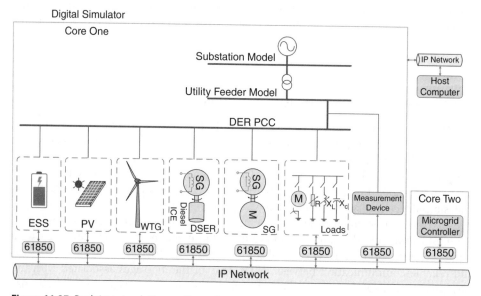

**Figure 11.27** Real time simulation configuration.

a Local Area Network (LAN). The GOOSE messages provide a real-time and reliable data exchange, based on a publisher/subscriber mechanism.

The real-time HIL set-up is comprised of one RTS, communicating using the IEC 61850 GOOSE Messaging protocol as shown in Figure 11.27. One simulator core is used to emulate the network and DERs, while the second core is used as a digital controller to implement the proposed microgrid controller as described in Section 11.1. The RTS provides an interface to publish and subscribe to GOOSE messages as per the Substation Configuration description Language (SCL) file defining IED configurations and communication. The RTS also provides a host computer interface whereby the user can monitor the real-time waveforms and pass new settings to the real-time model, if required.

# 12

# Conclusion

This book has presented design guidelines for microgrids. These guidelines consist of specification requirements, design criteria, recommendations, and sample applications for the stakeholder/client in order to comprehend the technologies, limitations, tradeoffs, and potential costs and benefits of implementing a microgrid. The guidelines take into account diverse applications of microgrids including urban, mining, campus, and remote communities. The development of this guideline was carried out using a systematic design methodology that comprises design criteria, modeling, simulations, economic and technical feasibility studies, and business-case analysis. In addition, the guidelines address the real-time operation of the microgrid (voltage and frequency control, islanding, and reconnection) as well as the energy management system (EMS) in islanded and grid-connected modes.

The design guidelines were categorized in two parts. The first part dealt with the research and development (R&D), which involved theoretical research work. The work included technical evaluation, economic and cost benefits, sizing storage systems, demand response (DR), power management, and control. The second part involved the implementation of selected potential technologies. Some demonstrative results were included in the book. Overall the book provides the design process and guidelines for a microgrid designer starting from design objectives selection and benchmark identification to the validation and testing of the designed microgrid for the selected design objectives.

## 12.1 Challenges and Methodologies

This section describes challenges and methodologies for each research topic within each of the three themes covered in this book.

### 12.1.1 Theme 1 – Operation, Control, and Protection of Smart Microgrids

As discussed earlier, this theme focused on operation, control, and protection of urban, rural, and remote microgrids in general and in particular on: (i) next-generation operational strategies of microgrid configurations; (ii) analytical tools for the analysis, design, and performance evaluation of concepts and technologies of microgrids; (iii) next-generation primary, secondary, and tertiary control strategies and the corresponding algorithms for implementation in digital platforms; (iv) novel adaptive

*Microgrid Planning and Design: A Concise Guide,* First Edition. Hassan Farhangi and Geza Joos.
© 2019 John Wiley & Sons Ltd. Published 2019 by John Wiley & Sons Ltd.

protection strategies and algorithms and the corresponding infrastructure for inter- and intra-microgrid applications; (v) methodologies to address the issues that impede large scale grid-integration of renewable resources in the next-generation microgrids; (vi) the state-of-the-art monitoring, detection, diagnostics methods and algorithms for operation, control, and protection of microgrids; (vii) power-management strategies for grid-connected, islanded, and virtual power plant (VPP) modes of operation of microgrids; and (viii) identification, adaptation, and applications of information and communication technologies (ICTs) for realization of the tasks associated with the above objectives.

### 12.1.1.1  Topic 1.1 – Control, Operation, and Renewables for Remote Smart Microgrids

The main objectives of Topic 1.1 were to research, identify, and develop robust control and protection strategies and algorithms, and the enabling ICT for optimal supervisory control mechanisms and power management of the Canadian remote microgrids, with emphasis on the maximization of the integration level of renewable sources, such as wind, hydroelectricity, and solar photovoltaic (PV). In particular, these objectives were:

*Objective I.* To develop a robust and fault-tolerant control strategy that can reliably provide optimum utilization of renewables in remote microgrids that are subject to frequent disturbances and load changes. Several methods were developed to minimize the impact of the load on the control and frequency of an islanded microgrid with multiple electronically interfaced distributed energy resource (DER) units. The basic idea in the proposed methods has been to allow the cooperation of the DER units through the widely accepted droop-based control at low frequencies (i.e. corresponding to slow-varying disturbances), while the dynamic coupling of the DER units, with one another and with the load, is cut at high frequencies (i.e. corresponding to fast changes). The results of this topic showed that the aforementioned approach can be implemented through robust control techniques.

*Objective II.* To develop a supervisory control that can dispatch and maximize the depth of penetration of DER units, including renewables in remote microgrids. Thus, a secondary level, robust, multivariable, Linear-Matrix-Inequality-based, Proportional-Integral control strategy was proposed to guarantee the stability of the microgrid with regards to the constraints. The proposed method is capable of coping with communication delays and partial disruption. Research was also conducted on optimal sizing and dispatch of islanded microgrids, which led to a proposed single-objective optimal sizing approach for islanded microgrids to determine the optimal component sizes for the microgrid, such that the life-cycle cost is minimized while a low loss of power supply probability (LPSP) is ensured. As wind speed and solar irradiation exhibit both diurnal and seasonal variations, the proposed algorithm takes advantages of the typical meteorological year-based chronological simulation and enumeration-based iterative techniques. Mathematical models were presented that consider the non-linear characteristics as well as the reactive power. The proposed sizing approach simultaneously provides the optimal component sizes as well as the power-management strategies.

*Objective III.* To identify and select reliable ICT and appropriate back-up strategies and algorithms to ensure reliability of supply. This objective was partially fulfilled, through the research toward objective II, which considered delays and disruption in communication.

*Challenges* Remote communities often have access to one or more renewable electrical energy resources, e.g. wind power, hydroelectricity, or solar PV power. Mainly due to geographical and/or economic reasons, most remote communities are either completely isolated from provincial and territorial grids, or connected to the power system via long and weak power lines passing through rough terrain. Such connections are often exposed to extreme weather conditions and consequently subject to frequent disruptions and downtimes. Therefore, the main source of electricity for remote communities has been off-grid diesel-generator units. In addition to the environmental concerns associated with diesel-generator units, fuel transportation imposes significant sustainability, cost, and logistical issues. Next-generation remote microgrids, based on maximum integration of renewables, are considered a potential solution to sustainable electrification of remote communities across the globe.

As compared with urban and rural microgrids, remote microgrids exhibit specific control and operational requirements and challenges. In some cases they can be viewed as a single electricity supply entity in which the main/dominant source is a set of diesel-generator units. In such a system the main objectives are: (i) to increase the depth of renewable resource units and provide them with the maximum power tracking control to reduce the use of diesel fuel to the minimum permissible limit; (ii) to ensure continuity of supply regardless of the type and the amount of renewables available at any time; (iii) to accommodate dominant loads, including motor loads, with non-conventional operational characteristics; and (iv) to provide control and enable continuity of operation under parameter excursions, e.g. voltage and frequency, beyond the limits are often imposed in conventional distribution systems.

On other occasions, the remote microgrid can be considered as an electrical subsystem, added to an existing electricity supply system which is relatively weak and often radial, and mainly supplied by diesel-generator units. In such a system, the function of the microgrid is to facilitate and maximize renewable integration levels, without jeopardizing the reliability of supply.

In an intelligent remote microgrid which mainly relies on ICT for control and protection, any potential malfunction and/or failure of the adopted ICT must not reduce the reliability and the continuity of electricity supply, and therefore redundancy and/or back-up strategies must be considered an integral part of remote smart microgrids.

*Methodology* Various types of remote power system configurations and the potential microgrid structures and their operational characteristics and requirements were researched and identified. Robust methods for accurate frequency and voltage control of remote microgrids were developed with consideration of the presence of single-phase and unbalanced three-phase loads and load uncertainties, to ensure power quality, enable optimal and stable operation, permit appropriate load sharing, and provide black-start capability. This also involved coordination of electronically interfaced generation; e.g. wind and solar PV, and conventional generation; e.g. diesel-, biomass-, and hydro-based generation. Research and development of robust and coordinated controls for real- and reactive-power outputs of the main and renewable resource units and storage devices allowed accommodation of the intrinsic nature and characteristics of renewables which could not be dispatched. The task of control was found to be equivalent to controlling a decentralized, Multi-Input/Multi-Output (MIMO), and time-varying system that used advanced control techniques. The optimal design and operational strategies of the remote microgrids were developed. This task also included R&D of the methodologies to determine the optimal types and ratings of the generators and storage devices, considering economic and environmental issues, based on the estimate of the loads types/profiles, the load growth, and potential renewable resources.

This undertaking also included the R&D of a supervisory control to monitor the remote microgrid in real time and determine the generation of dispatchable units. For non-dispatchable generators, such as wind turbines, the supervisory control had to define and adjust generation to ensure optimality. The balance of power could be maintained by energy storage devices. In view of the dispersion of generators and loads over the microgrid span, the supervisory control had to use ICT extensively.

Protection strategies and the required implementation technologies for remote microgrids may necessitate special consideration in terms of control and operational strategies of the renewable resources. Time-domain models would verify the developed control and protection strategies, including the required ICT. Digital hardware platforms may be used to facilitate digital hardware-based implementation of the developed control/protection algorithms for performance validation in a real-time, hardware-in-the-loop, simulation environment. This topic also considered the development of time-domain and frequency-domain models of the remote smart microgrid components for analysis, design, and performance evaluation of the envisioned control and supervisory strategies. The developed models were incorporated in the appropriate production-grade tools for analysis, synthesis, and performance evaluation.

### 12.1.1.2 Topic 1.2 – Distributed Control, Hybrid Control, and Power Management for Smart Microgrids

The main objectives of this topic were: (i) to conduct R&D of advanced control strategies and the corresponding algorithms for stable and optimal operation of urban and rural smart microgrids; (ii) to identify the ICT requirements and associated concepts and implementation technologies that were needed to realize the control methods/algorithms and decision-making supports; and (iii) to validate the developed algorithms, including detailed representation of the associated ICT, based on off-line digital time-domain simulation, real-time hardware-in-the-loop digital simulation environment, and beta-site tests at campus for pre-specified cases.

*Objective I.* Generalization of an ICT-based decentralized robust control method to accommodate simultaneously a large number of generation and storage units with significantly different time responses and operational constraints. Commercially available technologies related to this objective, particularly sensing and communication technologies, have evolved noticeably and as a result, the developed control strategies and algorithms are applicable to a wider class of application scenarios and are far more general than those originally planned.

*Objective II.* Provisions to guarantee adequate robustness limits of the decentralized controls subject to the smart microgrid's intrinsic unbalanced conditions and the wide range of change in the size, number, and characteristics of its generating units. This objective was to a large extent independent of technology and mostly based on well-formulated/understood mathematical approaches, i.e. design of the control and management algorithms.

*Objective III.* Development of a power-management system to provide optimal operation of the smart microgrid under various modes of operation with respect to different time frames, e.g. secondary and tertiary controls in the islanded mode. This objective was also modified during the course of the topic due to significant changes in market requirements for applications of microgrids. The modified objective met all the anticipated applications that have been considered as of now. The main outcome of this work includes three classes of power-management strategies and the corresponding algorithms for rural microgrids, urban microgrids, and the VPP mode of operation of urban microgrids.

*Challenges* In the conventional microgrid, the main focus has always been on the connection and operation of distributed generation (DG) units, mostly in the grid-connected mode. The control system mainly refers to the control of individual DG units to inject pre-specified real- and reactive-power in the system. Such controls are often augmented with feedbacks from voltage and frequency to provide some degree of coordination and load sharing among the DG units. If the microgrid is viewed as a unified power cell that requires operation in islanded mode, VPP mode, grid-connected modes, transition between the modes, and simultaneous inter- and intra-microgrid controls, then the conventional control approaches can neither provide optimal operation nor guarantee voltage/angle stability for the wide variety of operational scenarios that the microgrid is anticipated to accommodate. The challenges are numerous.

The envisioned rural and urban microgrids were understood to be inherently subject to a significant degree of imbalance which necessitated elaborate component models and specific consideration in the control design. Renewable sources, e.g. wind and solar PV units, which were expected to have a high depth of penetration in microgrids, were mostly non-dispatchable by nature. The DER units within a microgrid have noticeably different control characteristics and time responses. The controls of electronically integrated DER units were embedded and dynamically interacted with multiple protection functions. Furthermore, the electronically integrated DER and loads did not have the same type of inertial response as the directly coupled rotating generators in a conventional power system, where large inertia and slow dynamics exhibit significant stabilizing effect allowing more time for corrective control actions to be taken in cases of disturbances in the systems. Furthermore, a significant proportion of the electronically integrated loads, e.g. motor drives, plug-in electric/hybrid cars, computers, and appliances can introduce destabilizing or at least detrimental effects on the system. These factors impose additional challenges in terms of microgrid control and optimal operation.

Control requirements, priorities, and functions of a DER unit, for the host of transient scenarios that the DER was exposed to, varied drastically and required high levels of intelligence and up-to-date information to perform satisfactorily. The modes and nature of interaction phenomena and the patterns of behavior of DER units, particularly during the microgrid's autonomous mode of operation, were significantly different than those of the large, conventional turbine-generator units and were neither fully known nor comprehensively studied. The frequency of a microgrid was either subject to or required to have a wider range of variations during system dynamics as compared with that of the conventional interconnected power systems, particularly in the islanded mode.

A microgrid by default needs fast (ideally real-time) and accurate fault detection and identification due to its limited geographical span and the electrical proximity of its apparatus. Its relatively small size and the economic considerations dictate that a microgrid control system must be able to accommodate large variations in the number and characteristics of DER units under operation. The DER unit within a microgrid can have multiple control options that can be activated/de-activated to respond to specific operational conditions. The microgrid control system must be able to appropriately select the optimal control mode of the DER units and provide switch over between the modes. These factors represented a major departure from the existing control concepts of the conventional large power systems and the slow relaying schemes currently used in distribution systems.

Another major challenge, which was not constrained to this topic and equally applied to all topics under Theme 1, was the lack of appropriate component mathematical models, analytical tools, design tools, and performance evaluation media for the smart microgrid. The available analysis, design, simulation, and performance evaluation tools that were widely used for large, interconnect power systems mostly were either not directly applicable to the microgrid or could not accurately capture the inherently different performance characteristics of the microgrid. The main reasons were: (i) wide variety of new apparatus and components, particularly power electronic-based apparatus in the microgrid; (ii) presence of single-phase generation, loads, and lateral lines in the microgrid, and consequently high degree of imbalance during steady-state and dynamics of microgrids; (iii) significantly different operational modes of the microgrid as compared with the conventional large power systems; (iv) the need for frequent transition from one mode of operation to another mode with different operational characteristics and control/protection requirements; and (v) frequent and wide range of variation in the location, magnitude, and characteristics of generation and load under normal operating scenarios.

*Methodology* The methodology for this topic divided the smart microgrid control problem into three distinct sub-problems associated with (i) grid-connected mode of operation; (ii) autonomous or islanded mode of operation; and (iii) transition between these two modes. The grid-connected mode of operation was further subdivided into two groups, namely the conventional mode and the VPP mode. The first task was to identify and establish the structure, apparatus, and the operational characteristic and specifications for each mode, e.g. requirements for the presence of storage media and the maximum depth of penetration of non-dispatchable units. The next task was to identify, prioritize, and formulate the technical issues associated with the control, integrated control-protection, and operation of urban and rural microgrid configurations in the autonomous mode, grid-connected mode, VPP mode, and transition between the modes. Analytical tools for control design, such as linear dynamic models, were researched and developed to represent unbalanced conditions, lack of stiff source for synchronization, and large frequency variations. Signal processing and system identification methods were developed to extract the required control information for control design. Auxiliary controllers needed to be integrated and coordinated to provide inter- and intra-microgrid performance requirements. In this context, applications of phasor measurement units (PMUs) within the smart microgrid and for coordination and VPP mode of operation of multiple microgrids had to be researched and evaluated. This task required development of methodologies for phasor measurements which could: (i) accommodate large and fast frequency/angle excursions (as compared with those of conventional power systems); and (ii) enable digital implementation without the need for extensive computational hardware resources.

The then-existing control strategies that had been proposed for microgrid applications were intended: (i) mainly for real- and reactive–power injection; and (ii) limited frequency/voltage control and power sharing. They had been developed mainly for grid-connected mode and pre-specified islanded operating conditions. These methods suffered from one or more of the following limitations, weaknesses, and drawbacks: (i) sluggish performance of the controller and thereby they often violates the existing operational standards requirements; (ii) lack of adequate robustness and inability to accommodate various microgrid uncertainties; (iii) failure to maintain the stability and/or performance requirements of the system after parameter variations; (iv) dependency on a specific microgrid configuration; (v) the need for high-bandwidth and expensive communication links; (vi) lack of a back-up control scheme in case of communication failure; and (vii) the need for a central controller that required modifications to the microgrid control strategy and/or parameters after microgrid topological changes.

This topic devised a comprehensive control approach that addressed most of the above drawbacks of the aforementioned control methods. The control scheme proved to be high-performance, MIMO, robust, decentralized (MRD) control strategy, based on a generalization of a novel robust servomechanism controller that accommodated structural and parametric uncertainties based on non-conservative robustness constraints. The generalization was based on the use of ICT to guarantee robust performance of multiple generation and storage units within the smart microgrid during: (i) grid-connected mode, (ii) islanded mode; (iii) VPP mode; and (iv) transition between the modes. The MRD control scheme: (i) proved the need for communicating time-varying signals to a distant control system which required high-bandwidth communication link; (ii) eliminated the interdependency of local controllers; and (iii) reduced computational burden of a high order system, i.e. the smart microgrid, by splitting the system into multiple virtual subsystems and implementing the controller in a decentralized manner. The control system was structured as a central supervisory unity and an aggregate of number of decentralized local controllers corresponding to each generation/storage unit. This structure constituted a Hybrid controller.

The generation/storage units within the smart microgrid are typically categorized into two groups of dominant and smaller units. The dominant units, depending on the topology and size of the microgrid, are primarily intended to regulate the voltage of the system. The central controller determines reference set-points to achieve the desired microgrid voltage profile. The voltage regulating units also take the primary role of stabilizing the system during transients. Smaller units are required to maintain the generation and demand of the microgrid, and the central controller determines their active- and reactive-power set-points. The local controllers ensure the corresponding units track their set-points regardless of the disturbances and changes experienced by the microgrid. A back-up control scheme, which must be independent from communication links (such as droop-based methods), had to be devised to undertake the control task in case of communication failure. The control system also addressed the secondary frequency-voltage control and the tertiary frequency and/or voltage control of the smart microgrid. This task was carried out based on augmenting the smart microgrid control system with an "energy function" based computational engine to provide real-time secondary and tertiary control set-points for the corresponding voltage-frequency controls.

Smart microgrids are also typically equipped with a power-management strategy to ensure optimal operation and to accommodate: (i) intentional/accidental transitions between grid-connected and autonomous modes; (ii) islanding events within the microgrid; (iii) feeder transfer; (iv) faults, adaptive protection, and topological changes; (v) load or generation sudden changes; (vi) gradual but substantial change in load/generation; e.g. due to electric vehicle (EV)/plug-in hybrid electric vehicle (PHEV) units; and (vii) compliance with the market signals, fuel status, and the state of the charge in battery storage units. The power-management system requires real-time information regarding: (i) operational conditions, limits, status of both the internal microgrid controls and the external controllers; (ii) status of dominant generation units, major non-dispatchable units, and large storage units; (iii) EV/PHEV charging/discharging status; (iv) market signals, weather conditions, statistical pattern of load variations; and (v) sensors, monitoring devices, and automated meters. The power-management engine is required to process the information and perform real-time optimization to: (i) assign control priority/tasks to pre-specified apparatus; (ii) override specific control actions; (iii) disable/enable specific controllers; (iv) specify intentional circuit reconfigurations/changes; (v) determine planned islanding for specific DER units, portion, or all of the microgrid; and (vi) coordinate the internal and the VPP mode controllers.

The conceptual basis and mathematical formulations for the control and operation of a cluster of urban and rural microgrids, interconnected through a utility network, in which each microgrid operates as an intelligent unit were researched and developed based on the "intelligent agents" and "Autonomous Systems" concepts, which enabled real-time response to the market signals within the long-term dynamics.

Moreover, time-domain models were developed, using production-grade software tools to investigate dynamic and steady-state performances of the control strategies and to evaluate algorithms and the associated ICT. Digital hardware platforms for hardware-based implementation of the selected control/protection algorithms were also developed for performance validation in a real-time, hardware-in-the-loop, simulation environment. The Beta site microgrid at the campus was used for experimental verification of the selected control cases, in collaboration with the partner industries.

### 12.1.1.3   Topic 1.3 – Status Monitoring, Disturbance Detection, Diagnostics, and Protection for Smart Microgrids

The main objectives of this topic were to research, develop, and demonstrate protection strategies for urban and rural microgrids, including the required R&D of disturbance detection, status monitoring, and diagnostics to enable reliable protection under the envisioned operational modes of Topic 1.2 under all required conditions.

*Objective I.* Development of sensory methods and identification of the corresponding technologies for monitoring the smart microgrid, and the development of computational strategies for real-time extraction of the information for protection. This work led to two research achievements: (i) a state-estimation approach to provide improved visibility for the monitoring and protection of microgrids (involving a novel formulation of the state-estimation problem tailored to microgrids). (ii) a general purpose disturbance-detection method for both microgrids and interconnected systems. The new concept utilized in this work is the use of automatic and "soft" thresholds for disturbance detection. It is specially designed for detecting incipient faults.

*Objective II*. Development of an adaptive protection strategy that can reliably identify fault attributes in real time, subject to a wide variety of smart microgrid configurations, topological changes, and operational modes. The main outcome of this work was the proposition of a "decoupled" approach to deal with the protection coordination problem in microgrids. The idea is to limit the fault current contributions of distributed generators so that they cannot interfere with microgrid protection. For this purpose, a field discharging scheme to restrict the fault currents from synchronous DGs was developed. Effectiveness of the scheme has been verified by simulation studies. Design methods have also been developed. The impacts of inverter-based DGs on fuse-recloser coordination have also been studied. A novel control scheme that can limit the inverter fault current output has been proposed and verified. The fault current contributions of induction generators have also been investigated. It was found that this type of DG does not interfere with microgrid protection. Such DGs discharge fault current too fast for the protection to respond.

*Objective III*. Coordination and integration of the smart microgrid adaptive protection strategy with the fast controls of electronically interfaced DER units. Research was conducted in the areas of islanding protection, synchronization, and incipient fault detection, in view that protection coordination issues can be solved using the proposed "decoupled" approach. To this end, various islanding detection techniques were investigated. A power-line-signaling-based islanding detection technique proposed some years ago was found to be a superior method. As a result, new methods were not explored. A considerable amount of research effort was dedicated to the area of synchronization, and a novel, open-loop synchronization method for microgrids is proposed. This method is based on the idea of reducing switching transients associated synchronization, such as impedance insertion and point-on-wave closing. Research results have shown that the impedance insertion is the best candidate. An open-loop synchronization scheme has been developed based on this idea.

**Additional Work** An emerging issue of microgrid protection is the low voltage ride through (LVRT) capability of inverter-based DGs. Understanding this capability is very important for protection design. Therefore, this subject was also explored, which in turn led to the development of an analytical method to evaluate the performance of several key LVRT control schemes and new proposed methods to improve LVRT performances of inverter-based DGs. A stochastic simulation platform for modeling and simulating residential microgrids over a 24 hour period was also developed (and used to support the research conducted under objective I).

**Challenges** Generally speaking, protection concepts applicable to microgrids have been conventionally treated as an intrusion in the conventional protection strategy of the radial distribution systems and the relevant issues have neither been adequately investigated nor conceptually resolved. However, the recent envisioned operational modes, structure and technical features, and economic benefits of microgrids have necessitated serious consideration of new protection strategies, algorithms, and technologies to take full advantage of microgrid's features and benefits. The protection of microgrids, as compared with that of the transmission systems, is conceptually more complicated, and the tasks are further compounded by the fact that a microgrid is required to accommodate sound operation under (i) significantly different conditions, e.g. grid-connected and islanded modes; (ii) a wide variety of different types and number of generation units with significantly different transient characteristics; (iii) a high degree of imbalance in the lines, generation, and loads that can obscure fault and disturbance signatures; (iv) priority for continuity of supply and high power quality for pre-specified loads, even under a wide range of disturbances and load/generation energization and de-energization conditions; (v) transition to the autonomous mode and re-synchronization to the main grid; (vi) change of control strategies of DER units, e.g. the change from P-Q control strategy to voltage-frequency control strategy of electronically coupled DG units; (vii) the need to accommodate bidirectional power flow in any line segment; and (viii) the lack of clear guidelines, standards, and operational experience to define permissible range of variations and limits of parameters and variables.

Furthermore, in terms of protection issues, urban and rural microgrids typically exhibit specific challenges due to their intrinsically different configurations and characteristics, and as a result of the main grid specific conditions. An urban microgrid, which is often connected to a "strong" main grid with significant level of fault current contribution, has a relatively short electrical length, includes both underground cables and overhead lines, and can have an interface to the main grid at multiple connection points, e.g. the secondary distribution systems in the downtown core of large cities. In contrast, a rural microgrid is often connected to a fairly weak host grid which spans a significant geographical distance, primarily includes overhead lines and is subject to harsh weather conditions – e.g. storms, icing, and lightning – includes very long single-phase laterals, adopts multiple grounded neutral wires, and normally operates under significant unbalanced load conditions, including a large percentage of motor loads.

*Methodology* The existing distribution systems protection strategies, based on (i) pre-specified power flow direction; (ii) fixed relay settings; and (iii) relatively slow response, are not applicable to smart microgrids. The reasons are that in a microgrid: (i) successful operation of the control and power management relies on fast detection and isolation of faults/disturbances; (ii) the protection strategy must accommodate intentional and accidental topological changes without disrupting the system; (iii) relay settings must accommodate changes in the number, location, and status of generation and storage units, EV, and PHEV units and loads; and (iv) substantial imbalanced and shift of load/generation among phases must be accommodated. As such, Topic 1.3 focused on the concept of adaptive protection for smart microgrids, based on the use of: (i) multilayer and peer-to-peer communications; (ii) redundant signals from sensory, monitoring, and automated meters; (iii) centralized and distributed computational resources and logic elements; and (iv) distributed intelligent agents. The understanding is that the protection method must accommodate specific and peculiar distribution system characteristics; e.g. (i) three-wire and four-wire configurations; (ii) single-ground and/or multi-grounded lines; and (iii) various transformer connection types and structures. In that regard, the notion of learning-by-doing, based on the application of neutral network methods was entertained, to enhance and/or facilitate adaptiveness and performance of the envisioned protection approach.

The methodology also involved benchmark studies of systems that reflected intrinsic steady-state, dynamic, and transient characteristics of urban and rural microgrids, including control requirements, regulatory constraints and issues. Studies were performed on the benchmark system to identify and evaluate worst case scenarios, operational and parameter limits, regulatory and standard constraints and/or shortcomings. This information was used to develop analytical tools for on-line analysis of the microgrids for real-time monitoring and status evaluation based on information from distributed detection nodes and automated metering. This helped to diagnose stressed conditions for pre-processing the adaptive protection actions and decision making.

Strategies and algorithms were also researched and developed in order to specify corresponding technologies that detected and discriminated between fault scenarios and permitted disturbances, e.g. islanding, synchronization, DER energization and de-energization events, and sudden changes in the microgrid operational conditions, e.g. in response to market signals. Strategies and corresponding algorithms helped to identify the required ICT-based technologies needed to accommodate and process the outcomes of the above in either centralized, decentralized, or hybrid structures. Protection strategies that were based on local measurements and monitoring signals and auxiliary information proved to provide adaptive protection for microgrids. It was deemed important to evaluate the development of previous models based on time-domain simulation in production-grade software platforms, using the identified benchmark systems, and then identify beta sites in collaboration with the utility industry to validate findings on a beta-site microgrid.

### 12.1.1.4 Topic 1.4 – Operational Strategies and Storage Technologies to Address Barriers for Very High Penetration of DG Units in Smart Microgrids

The renewable energy penetration limit in a given distribution system is typically dictated by technical considerations, namely by thermal limits, related to the current carrying capability of the infrastructure, and performance requirements (generally dictated by standards). The former is essentially determined by the worst-case scenario, whereas the latter requires knowledge of the operating behavior of the system, including voltage profile (steady-state, flicker), frequency (if configured to operate in island mode), and reliability. To this end, modeling of the load, the DERs, distribution network operation, and their combined characteristics were required to ascertain how these requirements would be impacted, and consequently the penetration limit of renewables.

*Objective I.* Research and develop energy-management strategies and identify barriers associated with large number of renewable DGs. Work on this objective led to the identification of the microgrid controller metrics associated with grid-connected and islanded microgrids, for both urban and rural contexts. The topic also established a set of microgrid controller objectives and constraints to address the operational technical issues, taking into consideration the utility grid codes and standards.

*Objective II.* Establish performance merits considering cost optimization and provide evaluation based on simulation test cases. This topic developed off-line and real-time simulation models of the utility microgrid to evaluate the performance of a microgrid controller: (i) models developed for the DERs available and their associated local control loops; (ii) functions to evaluate the microgrid controller performance metrics; and (iii) modeling of the real-time microgrid controller EMS. In addition, the real-time hardware-in-the-loop testing platform was developed for microgrid control validation on the utility microgrid.

*Objective III.* Evaluate storage technologies, modes of operations, ancillary service functions and values. This research led to the development of: (i) a knowledge-based expert system (KBES) for the scheduling of an energy storage system (ESS) installed in a wind–diesel isolated power system; (ii) a novel isochronous control strategy for coordination of DERs, including storage, in an islanded microgrid; and (iii) a systematic approach to determine the optimized size of an ESS for energy cost reduction in a system with a high penetration of variable generation.

*Objective IV*. Research and develop strategies and algorithms to address barriers for penetration of large number of renewable based on storage technologies and ICT. Work on this objective led to a proposed KBES for the scheduling of an ESS installed in a wind–diesel isolated power system, and a methodology of formulating a multi-objective optimization (MOO) for a microgrid controller. Benefits achieved include minimized energy costs, reduction in peak power, power smoothing, greenhouse gas emission reduction, and increased reliability of service.

*Objective V*. Evaluate the performance of proposed strategies based on computer simulation test cases, and validate the storage and the ICT performance in the campus beta site. This topic led to a proposed implementation of the MOO for microgrid controller on the campus microgrid system (solar PV + battery + load).

**Challenge**  While technical limits impose hard constraints on the limit of renewables that can be integrated, economic considerations can lead to favoring one operating approach over others. Issues such as the time varying nature of the price of energy delivered or consumed by the microgrid, the way in which the infrastructure is used (such as overly frequent tap-changer operations for in-line regulators), and efficiency of the energy delivery process all need to be captured in an operational strategy.

An important characteristic that frequently impacts microgrids' performance with high levels of renewable energy is the myriad sources of uncertainty. Energy-management approaches that fail to capture this part of the problem will fall short of expectations. These sources include, as a minimum: the stochastic nature of the energy sources (renewables, controllable loads, and storage), the configuration of the network, and the duration of the microgrid in islanded mode versus grid-connected mode.

**Methodology**  An approach for optimal energy management of the microgrid was defined by formulating it as a mixed-integer, MOO problem. This approach required definition of the objectives of the microgrid, different operating modes, random variables describing the uncertainties, and models and constraints of the DER and the power-system infrastructure. The DER should include local generation, combined heat and power (CHP), electrical and thermal storage, and demand response resources.

To illustrate application of the methodology, one or more benchmark systems were used, using the base system and various scenarios of DER integration. This would permit comparison of microgrid performance, with and without appropriate EMSs and to the case of the system in traditional system operation. Representative data were obtained from a database common to all network partners of renewable energy production, load data, and power-system outage probabilities. This required coordination with other topics within Theme 1 and from the other Themes. These benchmarks systems was then used to demonstrate proof-of-concept, using quantifiable performance metrics.

This research relied on optimization tools such as General Algebraic Modeling System (GAMS) and publicly available tools, such as the Distributed Energy Resources Customer Adoption Model (DER-CAM) to develop the EMS. The system was simulated over at least a year of hourly data and was validated using its implementation in MATLAB™.

## 12.1.2 Theme 2: Smart Microgrid Planning, Optimization, and Regulatory Issues

This theme focused on issues related to the planning, economic and technical justifications of the creation of a smart microgrid, interaction with the main grid, including the utility regulatory requirements for connecting a microgrid to the main grid, the energy and supply-security considerations related to integrating a number of microgrids to the main grid, microgrid energy management, demand response and metering requirements and integration design guidelines and performance metrics.

### 12.1.2.1 Topic 2.1 Cost-Benefits Framework – Secondary Benefits and Ancillary Services

This topic looked into the general framework for the cost-benefits analysis and justification of creation of a smart microgrid. It also defined the grid interconnection requirements, from the microgrid manager and distribution system operator perspectives. Increasing the level of complexity of a given system was justified by the new functionalities or improved performance inherent in advanced smart microgrids. From an economic perspective, this line of argumentation was quantified by attaching monetary values to the cost of the additional complexity, weighed with the monetized benefits that were realized through these enhancements. For utilities and business owners, this analysis was required to develop the business case for a given technology, such that its integration could be justified to stakeholders or a regulatory body. This topic dealt specifically with the cost-benefit framework for smart microgrids and its feasibility as a new distribution system technology. It also took into account additional devices, such as storage, or power-management mechanism, such as demand-side management, which may be required to balance generation and load, particularly for variable generation such as that based on renewable resources.

*Objective I.* Establish the list of benefits and the framework for quantifying direct, energy supply benefits, taking into account utility interconnection requirements. Establish the principles of cost-benefit analysis for microgrids, including all benefits that may not be currently valued in markets, or where direct economic value may not represent the full benefit microgrids can provide to stakeholders. This work took into consideration multiple stakeholders and ownership models, and implemented the Use Case approach. A methodology was developed to evaluate key benefits, including: reliability improvement, ancillary service provision, investment deferral resulting from both peak load reduction, ancillary service provision, and Green House Gases (GHG) emissions reduction.

*Objective II.* Develop the methodology for quantifying the monetary value of the direct benefits. A methodology was developed to assess the impacts on the financial flows in an isolated remote community and mining microgrids of integrating wind, solar, energy storage and demand response technologies.

*Objective III.* Develop a framework for implementing and quantifying ancillary services. A framework and methodology and EXCEL spreadsheet-based analysis tools were developed to evaluate the benefits of providing ancillary services and applied the methodology to the frequency regulation service.

*Objective IV*. Planning and optimization approaches to maximize microgrid benefits. Through this research, the topic team developed: (i) DER and microgrid economic dispatch approaches and techniques for the operation of diesel-powered isolated remote communities integrating renewable resources; (ii) DER models and data required to represent wind, solar, and storage installations in microgrids in the planning stage; and (iii) EXCEL spreadsheet-based analysis software tools to evaluate the business case for grid-connected microgrids providing ancillary services. These tools represent an alternative to commercial analysis software DER-CAM, RETScreen, and HOMER, and were validated using the above.

*Objective V*. Application of the methodology to microgrid demonstrations in commercial, industrial, and remote community settings. This research produced a business case for a microgrid system and suggested alternative planning solutions to justify the economic operation of the campus Open Access to Sustainable Intermittent Sources (OASIS) system. A business case was also developed for each of the following: (i) a mining microgrid in Quebec; (ii) optimal planning of an urban microgrid in Calgary; and (iii) the use of Compressed Air Energy Storage (CAES) in combination with wind energy for remote communities.

*Challenge* Microgrids are often touted as a technology that can improve local system reliability, aid in the integration of renewable energy resources, and lead to enhanced power quality. However, to facilitate the additional control and operating modes associated with a microgrid, additional equipment is needed. The cost associated with this infrastructure can be quantified once the necessary elements are identified. Contrarily, there may be certain costs, such as those associated with changes to operating protocol, training, and new safety requirements, which may be more difficult to translate into a dollar value. Many of the benefits of a microgrid are also not tangible and consequently require additional considerations to monetize. One of the most important benefits of microgrids, reliability improvement, is poorly defined for a local distribution system; generally it is defined across a service territory. Furthermore, there is little consensus as to how improvements in power system reliability indices (e.g. the System Average Interruption Duration Index, [SAIDI] and the System Average Interruption Frequency Index [SAIFI]) can be monetized. Ancillary services are another possible benefit, yet ancillary service markets are generally not in place for the transmission system; at the distribution level they are non-existent. Thus, many of the benefits will need to be carefully discussed with industry partners to ascertain whether the assumptions made are reasonable. In some cases, new performance indices will need to be defined and debated among Network members. Once the issues of costs and benefits are well defined, the business case will then need to be developed and evaluated. To this end, some type of base case that can be used as a point-of-reference had to be defined. This was the case of the status quo; i.e. the distribution system prior to integration of DERs and microgrids, but this again needed to be debated by partners within the Network. A final looming question was: how can the assumptions made in the cost-benefit framework be tested? In other words, how can the resulting costs and benefits be checked to see whether they match those ultimately observed by the true system?

*Methodology*  All identifiable costs and benefits (including GHGs) associated with a smart microgrid were first enumerated and classified according to the difficulty with which they can be monetized. As mentioned previously, this required the definition of new performance metrics and assumptions on the value of a unit improvement of these measures. Because microgrids greatly impact the way the system is operated, current operating practices and system constraints (both of the distribution system and of the individual DER) would have to be well understood and adequately modeled. The cost-benefit framework would then have to be applied to a series of case studies (types of network, types and amount of local generation, existing operating approach). The feasibility of the microgrid concept had to be assessed for each of the cases. Furthermore, sensitivity analyses were performed to determine how the business case changed according to different economic scenarios (energy prices, fixed costs of microgrid technologies, financing models). Where possible, the analysis attempted to rely on tools for economic evaluation of renewable energy topics (RETScreen, HOMER) to perform the economic analysis, including sensitivity analyses. The results were then compared with operating data from the other topics within this Theme to determine whether the expected benefits were in fact expressed and to the degree predicted by the economic analysis.

*Results*  The results of the research answered the following questions: given today's prices and the best estimates on extrapolation to future prices, under what conditions does the microgrid concept make economic sense? Is it application dependent (e.g. downtown network, commercial district, industrial plant, rural system)? Is it DER dependent (e.g. technology and penetration level)? The business case should be able to quantify the value of microgrid implementation using predefined performance metrics. The most important parameters influencing its feasibility and economic viability were important conclusions that helped to quickly identify viable industrial and commercial topics.

### 12.1.2.2  Topic 2.2 Energy and Supply Security Considerations

*Challenge*  This topic dealt with integrating multiple urban and rural microgrids in the interconnected power system. This topic first evaluates and quantifies the impacts of microgrid integration on the host interconnected power system in terms of: (i) technical performance; (ii) reliability of supply; (iii) economical aspects; and (iv) potential infrastructure needs, in particular the data communication supervisory network and related information technology. Based on the results of the analyses, this topic also provides a host of remedial solutions to address the technical issues, the market requirement, and the standards and regulatory aspects of a power system that incorporates multiple microgrids.

Conventional urban and rural microgrids mainly deal with the "connection" and "operation" of DER units, primarily to accommodate the operational, control, and protection philosophy of the host distribution and/or the power system. Thus, minimal impact on the upstream system is anticipated. The smart microgrid, based on the use of data communication network, intelligent sensors, automated meters, and centralized supervisory controls, presents a microgrid as a cell with the intra-cell interaction characteristics. The intra-call interaction is anticipated to be of a controlled nature to assist the host upstream system, and have minimal interference and adverse impact on the overall system operational characteristics. However, the challenge is that as the number of microgrids integrated within a power system increases:

- The mutual dynamic interaction phenomena between the cluster of microgrids and the host power system can exhibit detrimental effects on the host power system in terms of reliability of power delivery, transient performance, optimality of power flow, protection requirements, power quality, and market aspects.
- The interaction phenomena among the microgrids, through the host power system, can manifest itself as a limit to optimal utilization of the microgrid features and assets, and even render the whole concept of microgrids either trivial or even technically and economically unattractive.
- The interaction phenomena (i) among the microgrids and (ii) within the cluster of microgrids and the host network can introduce technical barriers to the successful implementation of "active distribution network" and "smart grid" features, concepts, and technologies.

The interaction phenomena cover a wide frequency span and experience in different time frames; i.e. milli-second up to multi-minute and even the steady-state, and thus require assessment and provision for the overall system protection, primary, secondary, and even tertiary controls.

The focus of this topic were: (i) to research, identify and quantify the simultaneous impact of multiple urban and rural microgrids on the host power system; (ii) to develop the models, analytical methods, and the software/hardware-based simulation tools for the investigations; (iii) to research and evaluate the ramifications of multiple micro-grids on the implementation of "active distribution system" and "smart grid" concepts; (iv) to identify, research, and develop the ICT and concepts that are needed to realize the control methods/algorithms and decision-making supports; and (v) to validate the developments based on digital time-domain (real-time and/or off-line) simulation test cases, and beta-site tests for a set of specific cases.

These problems have neither been fully understood nor systematically investigated, and even the models and the analytical tools have not been developed. It should be noted that the solution to the above issues must simultaneously address the technical aspects, the standard requirements and the regulatory and policy needs.

*Objective I.* Quantify the impact of a large penetration of microgrids on the steady state, dynamic, and transient performance of the interconnected power systems. This objective exclusively dealt with technical aspects of microgrids and their associated generation units, i.e. dynamical behavior. The main outcomes of this work included: (i) a comprehensive, quantitative guideline to predict behavioral impact of microgrids on their host power system during steady-state, small-signal dynamics, and large-signal transients; (ii) a comprehensive, quantitative guideline to identify suitable control methods to prevent undesirable behavior subsequent to accidental transients within microgrids; and (iii) a list and categorization of transient phenomena that can cause microgrid variables to reach critical values and potentially result in tripping and de-activation of DERs.

*Objective II.* Determining the potential violations and/or infringements of the regula-tory requirements and standards, as well as guidelines that need to be established for a large penetration of microgrids. This objective proved to be extremely difficult to achieve due to the fact that regulatory requirements, acceptable guidelines, and the connection limits were and are highly dependent on: (i) the jurisdiction under consideration; (ii) unit and microgrid size; (iii) unit type; and (iv) the host system structure and operational characteristics. Therefore, the scope of the objective was restricted to the Ontario power grid and further divided into the guidelines for high penetration levels of distributed resources in urban microgrids and rural microgrids. The outcome is a set of guidelines to meet both steady-state and transient requirements for microgrid and DER integration in Ontario.

*Objective III*. Developing supervisory control and power-management strategies and algorithms to enable partial and/or full autonomy of microgrids and to minimize adverse impacts within microgrids. This objective is the main criterion for operation of the microgrid in the VPP mode both in terms of system dynamics and response to market signals by dividing the time frame of the response of a microgrid to external disturbances/transients into four distinct time frames, every fifteen minutes after a disturbance/change occurs in the microgrid. The first time frame covered the first five cycles (of 60 Hz), the second time frame covered the next 45 cycles, the third time frame covered up to two minutes and the fourth time frame covered up to 15 minutes. Whenever a disturbance occurred, the first time interval was initiated and the subsequent intervals were considered. If no disturbance encountered, then the fourth interval continued until a change/disturbance occurred. For each interval the system was analyzed separately, in terms of modeling and the analysis tool, and accordingly the control action was imposed by the supervisory control on the local controllers.

*Methodology*   The methodology to achieve the topic objectives is as follows:

- Identify and classify the technical problems, potential violations and infringements of the existing standards and guidelines, new regulatory issues, and policy requirements associated with the integration of a large number of urban and rural microgrids in the power system. In the Canadian context, special attention should be paid to issues associated with the presence of multiple microgrids in a portion of a power system that is not highly meshed, and even may primarily rely on a single radial power transfer corridor.
- Research and develop the models and the corresponding analytical and simulation tools for: (i) the assessment of the various operational conditions; (ii) identification and quantification of potential technical, economic, regulatory and standard issues; (iii) synthesis of operational, control, and protection strategies and the corresponding algorithms; (iv) intelligent sensors meters, data communication networks, and information technology infrastructure; (v) advanced technologies; e.g. ESSs and power electronics-based apparatus such as the sub-cycle transfer switch; (vi) advanced operational concepts and strategies; e.g. active distribution system and smart-grid; and (vii) quantitative evaluation of the envisioned solutions.
- Assess and potentially modify/amend the strategies, methodologies, and the tools that have been traditionally considered as the pure territory of the large interconnected power system such as: (i) mid-term and long-terms stability analysis and enhancement; (ii) the secondary and the tertiary voltage/frequency control; and (iii) the optimal power flow (OPF) to represent the sub-transmission and distribution systems with the required details to reflect the impacts of smart microgrids.

- Research and develop supervisory networks based on the intelligent sensors and diagnostics algorithms, automated meters, communication infrastructure and information technologies for decision-making support, visualization, and coordinated operation of the main-grid and its embedded multiple smart microgrids. This R&D task will utilize state-of-the-art developments and usher new advances in synchronized phasor measurements strategies and technologies to provide monitoring, protection, and control of the main grid and its integrated multiple microgrids.
- Research and develop strategies and algorithms to maximize efficiency of the integrated main grid and the microgrids, subject to reliability related constraints, power-quality limits, and environmental requirements.
- Specify benchmark system(s) and develop time-domain, (real-time, and off-line) models, using production grade tools, to investigate and evaluate the developments of the abovementioned R&D results, considering the supervisory network and its communication infrastructure and the information technologies.
- Identify a beta site system, in collaboration with the utility industry, for experimental evaluation, verification, and demonstration of the selected R&D results.
- Specify the supervisory intelligent network and identify the corresponding technologies for secure and coordinated operation of a large power system that embeds multiple intelligent and conventional (rural and urban) microgrids.
- Provide the protection, control, and operational strategies and algorithms, based on the use of ICT, intelligent sensors, and automated meters to enable secure, reliable, and coordinated operation of the overall main grid and its embedded (urban and rural) microgrids.
- Address the required changes/modification in the standards/guidelines and identify the new policy and the regulatory issues associated with the high level of integration of microgrids in power systems.
- Provide models and the analytical tools to assess performance and enable systematic design of the envisioned strategies and algorithms.
- Provide case study results of benchmark systems to quantitatively access merits/limitations of strategies, algorithms, and technologies, considering data communication, intelligent sensing and monitoring technologies.
- Perform beta site tests on a selected set of scenarios to demonstrate the topic results.

### 12.1.2.3 Topic 2.3 Demand-Response Technologies and Strategies – Energy Management and Metering

This topic focused on developing demand-response strategies and technologies based on understanding the load profile of target loads through sensing and metering. Data collected through real-time measurements were used as the basis of energy-management decisions and approaches in line with the system's predefined energy-management attributes and requirements.

*Objective I*. Determine the environmental, economic, and social impacts of an increased deployment of microgrids, and their overall sustainability. This research developed new mathematical models for representation of emission characteristics of diesel generation units in isolated microgrids. These models were integrated within the EMS framework, different multi-objective-based microgrid EMS (MEMS) models were then formulated which simultaneously minimized the operating cost and pollutant emission costs. To highlight the effect of demand response on the pollutant emissions and microgrid operation, a constant energy demand shifting model was formulated for smart loads. To cope with the uncertainties, a model predictive control (MPC) technique was adapted.

*Objective II.* Designing energy-aware scheduling algorithms and determining the cost and benefits of such technologies to be implemented within microgrids. This work established a mathematical model of smart loads in DR schemes, which were integrated into centralized unit commitment (UC) with OPF-coupled EMS for isolated microgrids for optimal generation and peak load dispatch. The smart loads were modeled with a neural network (NN) load estimator as a function of the ambient temperature, time of day, time of use price, and the peak demand imposed by the microgrid operator. To develop the NN-based smart load estimator, realistic data from an actual energy hub management system was used for supervised training. Based on these, a novel MEMS framework based on a MPC approach was proposed, which yields optimal dispatch decisions of dispatchable generators, ESS, and peak demand for controllable loads, considering power flow and UC constraints simultaneously. The impact of DR on the microgrid operation with the proposed MEMS framework, was examined using the International Council on Large Electric Systems (CIGRE) benchmark system that includes DERs and renewables-based generation. The results showed the feasibility and benefits of the proposed models and approach.

*Objective III.* Propose algorithms for optimizing the internal microgrid load balancing capability taking into account impacts and constraints on the reliability and capacity of the main electric grid. A frequency-control mechanism was developed for an isolated/islanded microgrid through voltage regulation. The proposed scheme made use of the load voltage sensitivity to operating voltages and was adopted for various types of isolated microgrids. The proposed controller offered various advantages, such as allowing the integration of significant levels of intermittent renewable resources in isolated/islanded microgrids without the need for large ESSs, providing fast and smooth frequency regulation with no steady-state error, regardless of the generator control mechanism. The controller required no extra communication infrastructure, and only local voltage and frequency parameters were used as feedback. The performance of the controller was evaluated and validated through various simulation studies in the PSCAD/EMTDC (Power Systems Computer-Aided Design/Electromagnetic Transients including DC) software environment based on a realistic microgrid test system, using small-perturbation stability analysis to demonstrate the positive effect of the proposed controller in system damping.

**Challenge** EMS use advanced control and communication technologies to send signals to load clusters that are targeted during a critical or peak demand period. In microgrids, the demand responsive loads can be activated upon notification and are integrated into the optimization strategy. This information will be conveyed for distributed automation of the microgrid using appropriate communication protocols for the purposes of managing the microgrid during transitional steps between islanded and grid reconnection. Rapid and automatic demand-response schemes, coupled with on-site energy generation, need to be evaluated to quantify the range of energy savings from peak shaving and define additional provisions required to operate the system close to the margins while maintaining the highest power service reliability. The topic, therefore, considered optimal asset management through demand response, peak shaving, and outage restoration. In addition, it studied the attribution of benefits by assessing market pricing approaches for microgrid investors as a function of regional or nodal network constraints. The originality of this research was that in addition to developing novel algorithms and techniques for each of the topics, it drew on the expertise required to develop a comprehensive smart grid system.

*Methodology* The purpose of this topic was to evaluate the benefits and costs associated with the use of microgrids, as compared with conventional, large-scale grids. More specifically, the consequences (broadly defined, in terms of economic, social, and environmental impacts, accruing to a range of individuals and organizations within the particular jurisdiction of interest) were assessed through modeling and measurement means. Two issues were investigated: consequences arising from the fact that microgrids can respond quickly (and thus have particular "temporal implications"); consequences arising from the fact that microgrids can be located close to load centers (and thus have particular "locational implications"). Given that approach, the following studies were done: (i) a microgrid temporal consequences and peak-shaving study to investigate the consequences of reduced use of the centralized grid, as well as increased use of DG on meeting load demand. A general model to evaluate different situations was constructed. Inputs were determined and a selection of case studies made, both "conceptual" and "empirical," which may differ based on different carbon ratings of the systems. A sensitivity analyses were conducted leading to policy implications and other conclusions; and (ii) the microgrid locational consequences and nodal pricing study to investigate the consequences of reduced use of the centralized grid, as well as increased use of DG on energy pricing. A general model to evaluate different situations was constructed, with inputs determined, and a selection of case studies that were both "conceptual" and "empirical." Case studies differed across different carbon ratings of the systems. Sensitivity analyses were conducted. Policy implications and conclusions were drawn.

### 12.1.2.4 Topic 2.4: Integration Design Guidelines and Performance Metrics – Study Cases

This topic had the essential objective of reviewing and affirming integration design guidelines based on target performance requirements. The plan was to identify various existing component models, define various microgrid topologies and configurations and develop preliminary models based on microgrid size (MegaVolt-Ampere [MVA]), location and mix of generation (conventional/distributed/renewable/storage). Furthermore, it was needed to define the control/communication layer and determine modeling requirements, including technical aspects such as communication latency and spread, failure modes, and durations. The work involved constructing a highly detailed electromagnetic simulation of a small microgrid to be used for additional studies to determine adequacy of reduced order models, and stability study models, based on different operating scenarios. The developed models had to be refined through case studies and their validation for various scenarios such as: (i) integration of renewable energy resources into the microgrid, including the possibility of local storage; (ii) determining islanding and disturbance ride through effects for isolated microgrids; (iii) sharing resources between microgrids; and (iv) operation of microgrids under failure or degradation of communication systems.

*Challenge* Although much work has been carried out on the modeling of various components of the microgrid, such as renewable energy sources, relatively little has been done on the integration of these into models of the microgrid itself. One challenge is to define what components and what mix of components to include in the model. The microgrid's characteristic may be quite different depending on the source/load mix. One visualization paradigm is to consider the overall power network as a collection of microgrids connected by way of major transmission connections and also by suitable information exchange connections for control and protection information interchange. Modeling the grid in this form had not been reported in significant detail in literature, and hence was a new challenge to be addressed in this topic. Another challenge was to determine the level of detail necessary in modeling the microgrids for different types of topics. Excessively detailed models might be suitable for determining and rectifying problematic fast transients, but would be too cumbersome for larger system level studies and require reduced order models. The topic constructed models for microgrid components, connecting electric network layers, renewable generation equipment and simplified models showing essential control interactions between the microgrids. These models were used to construct several case study examples to validate or demonstrate approaches that were generated in the other themes. The topic was innovative because it considered an emerging and important concept in modern power systems; that of evolving the existing power system into several autonomous locally controlled microgrids incorporating new sources of generation. Such systems had never been comprehensively modeled in the past and there was a need for the development of models and benchmark test systems to gain experience with their operation.

*Methodology* The topic involved the following steps:

- Compile and study the available models for microgrid elements, including energy sources, loads, distribution and transmission systems, and local energy storage options (including equipment such as batteries and plug-in EVs).
- Define several different microgrid structures based on different types of energy sources, loads, and storage elements.
- Model a control and communication network of the microgrids.
- Define typical study requirements based on the aims of other topics in all three themes and then define the systems required for case studies, including benchmark systems.
- Determine a small, highly detailed microgrid model as a template and basis for developing reduced order models.
- Analyze modeling detail versus model detail tradeoffs.
- Develop models for the various study areas.
- Conduct complete sample case studies on various microgrid operating scenarios and contingencies.
- Validate models with field results from campus microgrid and the utility microgrid (as applicable).

### 12.1.3 Theme 3: Smart Microgrid Communication and Information Technologies

As discussed earlier, this theme focused on network architectures, capable of supporting the exchange of data and commands between various entities in a smart grid system.

In addition to various communication architectures suitable for such transactions, the topic was required to address the impact of communication systems on the control system dynamics, reliability, resiliency, and security of the network.

### 12.1.3.1 Topic 3.1 Universal Communication Infrastructure

The topic dealt with best practices for constructing wireless, wired, and hybrid communication systems; reliable and cost effective communication protocols, and robust authentication and encryption methods for data/command exchange between smart grid components. It also aimed at providing researchers and developers with the guidelines to accurately simulate and predict the performance of wireless infrastructure that may be deployed within smart microgrids. The topic assumed the following objectives:

*Objective I*. Evaluate and assess techniques for exchanging data and commands through hybrid networks and schemes for authentication, authorization, and access within smart microgrids.

*Objective II*. Propose and evaluate alternative techniques for exchanging data and commands through hybrid networks for authentication, authorization, and access within smart microgrids.

*Objective III*. Conduct interference studies and develop deployment guidelines for communication systems within smart microgrids.

*Objective VI*. Conduct interference studies and develop deployment guidelines for wide area environments.

*Challenges* The communications infrastructure that serves smart microgrids needs to be deployed as a hierarchy of networks, each of which may be realized using various wired and wireless technologies. At the lowest level, home area networks or HANs will connect devices and displays within a building to the smart meters (SMs) that connect the customer premises to the distribution grid. The SMs will relay real-time energy consumption data to the utility for use in accounting, billing, load forecasting, and outage detection while relaying information concerning pricing policies and possibly load reduction or load shedding commands to the customer. At intermediate levels, e.g. neighborhood area networks or NANs, the SMs and other sensors will connect to data aggregation units (DAUs) or similar that will be deployed throughout each neighborhood. At the higher levels, e.g. wide area networks or WANs, the DAUs will connect to base stations or nodes with direct connections to the utility's core network. At each level, the coverage and mutual interference, reliability and latency, and authentication and security requirements must be assessed and means devised to satisfy them.

Although the backbone infrastructure used by the Bulk Power System is very reliable and well-suited to serving the relatively limited number of nodes that lie at the substation level or above, scaling up such a system to serve the large number of network termination points or terminal nodes in a microgrid communications network would be prohibitively expensive. Meeting strict budget constraints will require that new wired and wireless technologies be developed and deployed. Using wireless infrastructure in smart microgrid applications is attractive because it can be rapidly deployed without incurring the effort associated with installing wires and cables and it is relatively immune to the type of physical damage to the cable plant that can render wired infrastructure unusable. Wired infrastructure, especially power line communications (PLCs), is attractive when connectivity must be supplied in in-building environments and confined spaces where providing wireless connectivity is difficult. Both technologies may suffer from range limitations, gaps in coverage and mutual interference that can significantly render them ineffective. Vendors and operators require clear guidelines and best practices for the deploying wireless and wired networks that will provide the desired performance at a reasonable cost with due consideration for regulatory guidelines.

The large number of network nodes, the large amount of data to be handled, and the relative complexity of the network topology places a relative premium on the efficiency of the protocols used to form networks, transfer data across gateways between networks, and move data from meters and sensors through data aggregation nodes to the servers that store the state of the system and the accounting and billing records. Link-level performance will affect reliability and throughput. Devising appropriate scheduling and admissions policies is essential if network performance is to effectively handle different types of traffic with differing quality-of-service requirements. At all levels (HANs, NANs, and WANs), robust authentication, authorization, and access control schemes must be developed with due regard for integrity, robustness, and efficiency.

*Methodology* The proposed research topic involves three major tasks. The first major set of tasks is to characterize wireless propagation in target environments. The measurement-based models so produced will form the basis for simulating and predicting the performance of wireless infrastructure under realistic conditions including assessment of the performance of alternative protocols and routing algorithms and the level of interference that will be generated by networks. The final step is to develop clear guidelines and best practices for deploying wireless networks in home area, neighborhood area and wide area environments.

The second major set of tasks will involve: (i) development of strategies for appropriately combining wired and wireless infrastructure in order to yield optimal coverage and performance under a variety of conditions; (ii) assessment and improvement of current techniques for exchanging data and commands through hybrid networks; and (iii) development of protocols and policies that will ensure optimal network performance, e.g. throughput, reliability, and latency, in both the upstream and downstream directions.

The third major set of tasks will involve: (i) assessment of schemes for authentication, authorization, and accounting; (ii) identification of shortcomings; and (iii) proposal of improved schemes that will overcome previous limitations.

### 12.1.3.2 Topic 3.2 Grid Integration Requirements, Standards, Codes, and Regulatory Considerations

The characteristics of different information types in smart microgrids were studied to establish quality-of-service parameters and to classify their dynamic quality-of-service requirements. In consideration of information characterization, emerging standards were studied to develop efficient transmission, information processing, and networking techniques and strategies suitable for a robust communications infrastructure that could support the integration of smart microgrids. In considering the grid integration requirements, standards, codes and regulatory issues, this topic has taken into account the relevant peculiarities of adopted protocols and the security of relevant infrastructure, distribution automation (DA), data extraction and organization. In particular, robust information transmission techniques were developed to enhance the performance of communication channels in the distribution system, data extraction and organization. Intelligent information-processing strategies for timely and reliable delivery of measurements and commands facilitated not only effective data collection but also efficient universal communication infrastructure.

*Objective I.* Identify optimum communication technologies for integration of smart microgrids as a function of required transactions. A study was completed on various potential communications technologies that could be used for HAN/Building Area Network (BAN), Industrial Area Network (IAN), NAN/Field Area Network (FAN), and WAN that form the communications infrastructure to effectively support the integration of intelligent micro smart-grids/microgrids. Those technologies included IEEE 802.15.4/Zigbee, IEEE 802.11/Wi-Fi, WiMAX, 3/4G Cellular, PLCs and other wire-line communications. The advantages and disadvantages of each technology in terms of transmission rate and coverage, maturity of standards, deployment and maintenance costs, etc., were identified in the context of communications infrastructure for microgrids in order to develop guidelines for technology integration. Additionally, the investigation of various information types in intelligent smart-grids/microgrids were investigated, as well as their communications traffic characteristics, to derive their quality-of-service parameters and to classify their quality-of-service requirements in terms of communications bandwidth, delay, and reliability. Based on the observations and conclusive results obtained, a communications infrastructure was suggested together with various communications technologies (wired/wireless) that could be used to enable current and future applications for smart-grids/microgrids.

*Objective II.* Identify standards for end-to-end messaging, command, and control among integrated microgrids. For this objective, interoperability issues were investigated in smart-grids/microgrids. The importance of interoperability standards for smart-grids/microgrids was highlighted. Various standards development organizations as well as alliances that have been working toward the development of interoperability standards were surveyed. In particular, the National Institute of Standards and Technology (NIST) and its activities on standards for smart-grids/microgrids were focused on. A collection of standards that have been reviewed by NIST to be applicable to microgrid integration were mentioned. Priority Action Plans (PAPs) that NIST has identified for developing and improving standards necessary to build an interoperable microgrid were also investigated. These PAPs attempt to fill gaps in standards (i.e. a standard extension or new standard is needed) and to resolve overlaps in standards (i.e. two complementary standards address some information that is in common but different for the same scope of application).

A number of representative standards dealing with smart metering, building automation, substation automation, DERs integration, and PHEV were studied. They include IEEE 2030, IEC 61850, DNP3, ANSI C12.18, ANSI C12.22, IEEE 1547, IEC 61850-7-420, IEC 61400-25, IEC 62351, and SAE J2293. Furthermore, the applicability of the IEC 61850 standard and PLCs were evaluated for Advanced Distribution Automation (ADA) applications in microgrids. First, data modeling/data mapping and communications architectures and protocols of the IEC 61850 standard were studied. The major advantage of this standard is that it can offer interoperability for devices from different manufactures. Therefore, it appears as the promising standard for ADA. Second, the advantages and limitations of PLC as the communications technology for ADA were investigated. A general framework for evaluating the bandwidth requirement for ADA has been presented. This study revealed that although PLC exhibits many advantages (e.g. low cost, fast deployment, high customer acceptance, etc.), it is still facing many technical challenges, including limited data rate, topology inflexibility, and communications loss due to power outage.

In addition to the survey on data communications standards for smart-grids/microgrids as planned, a tree diagram was developed summarizing key standards/technologies and their respective technical details as well as an evaluation on the applicability of the IEC 61850 standard and PLCs for ADA applications in microgrids.

*Objective III.* Determine efficient protocols for supporting distribution automation within and across integrated microgrids. A comprehensive survey was completed on smart-grid/microgrid applications and their data communications requirements (in terms of throughput, transmission reliability, and latency), referring to information provided in PAPs (NIST) and 2030 (IEEE) with considerations on existing and emerging smart-grid/microgrid applications/use cases. They then focused on the development and evaluation of efficient routing protocols and associated simulation tools for quality-of-service provisioning in NANs/FANs – the most important communications network segments.

A network simulator was developed that facilitated the study of the operation and performance of various wireless routing protocols for NANs. The simulator modeled the NAN consisting of a number of end-points including SMs installed at customer promises, and data aggregation points (DAPs) usually placed on top of poles or in substation areas. SMs collect customer's energy consumption information and route it to DAPs over single-hop or multi-hop paths. DAPs then forward information to central control centers. Besides, control information can be sent from control centers to SMs in the opposite direction. The two promising wireless routing protocols that could be used for NANs were implemented in the simulator. The first one was Greedy Perimeter Stateless Routing (GPSR), a simple yet efficient location-based protocol. This protocol was ideal due to its low complexity, low overheads, and reliability. The second one was Routing Protocol for Low Power and Lossy Networks (RPL), a state-of-the-art self-organizing protocol. One of the key features of RPL was its ability to capture wireless link dynamics to build up a Directed Acyclic Graph (DAG) which was used to route packets from sources to destinations over efficient paths.

Extensive sets of simulations have been performed to verify the operations of GPSR and RPL and to evaluate their performance in different network scenarios. System parameters (e.g. placement and density of SMs, number of SMs per DAP, traffic profiles, etc.) presented by NIST have been considered. A number of performance metrics have been measured, e.g. packet delivery ratio, end-to-end latency, and signaling overheads. GPSR and RPL were compared to identify their applicability is various smart-grid/microgrid scenarios.

This work took into consideration a real-life scenario – the Burwash Landing microgrid. Burwash Landing is situated on the western shore of Kluane Lake (Yukon) and is home to the Kluane First Nation (KFN). The operation and performance of the proposed wireless mesh networks with GPSR and RPL routing protocols was explored to support the communications for various microgrid applications in Burwash Landing.

*Challenge* Robust, reliable, and forward-looking strategies for the integration of smart microgrids need to facilitate the emergence and evolution of smart grids. Robust communication has been identified as the primary challenge to deploying automated, prompt, self-healing wide-area grid-control actions. The desired topology for an open communications infrastructure includes at a minimum a network of a heterogeneous wireline and wireless communications segments and data gathering and control sensor networks. Various activities/standards such as IEC61850, UCA2, IEEE P1777 have laid down initial foundations for further studies and the development of a robust communications infrastructure to effectively support smart grid integration. "Robust communications" implies "reliable," "timely," and "efficient" data exchanges required for effective measurement, monitoring, fault detection, control, and management functions in DERs. For example, wireless communications with a wide range of available and emerging technologies are very attractive for mobility, rapid installations and network re-configurations, easy maintenance, safety due to remote operations, low costs while their transmission reliability can be affected by random interference and fading. Internet using transport control protocol (TCP)/internet protocol (IP) based on best-effort service and re-transmission may not be adequate for delivery of time-sensitive or delay-constrained information in power-system-grid management, e.g. phasor measurement [71]. Furthermore, as another example, power/frequency disturbances tend to produce system-wide clues [72], and efficient data exchanging techniques and strategies regarding power/frequency measurements can offer opportunities to observe the propagation of the disturbance across the entire network in a timely manner for effective fault detection, prediction and isolation.

*Methodology* Suitable techniques and strategies need to be developed for the communications infrastructure to support integration of smart microgrids with special emphasis on its robustness, security, and ease of maintenance. To assess the communications requirements for grid integration, one needs to characterize the information in electricity distribution systems, specifically in smart microgrids, from a communications' vantage point. In general, the majority of information is related to measurements, and parameter settings, which can be classified as inputs and outputs, respectively, in a very simplistic view. However, such a simplistic classification does not represent the information importance and urgency that determine different service priorities and requirements. Furthermore, the optimal service priority and quality of a particular measurement may be dynamically changed according to global and local attributes. As an example, consider voltage measurement, which is a universal function across the distribution network. Within some range it can be considered normal and hence reported as expected sensory data, while in another range, it may be regarded as critical, and in need of high-priority services. As such, the characteristics of various information types in smart microgrids need to be studied, aiming at establishing their quality-of-service parameters and classifying their quality-of-service requirements. Here, a new notion of dynamic, context-aware quality-of-service needs to be introduced. On one hand, such studies need an in-depth understanding of measurement, monitoring, control, and management techniques and functions in DER. On the other hand, the analysis and mapping of quality-of-service parameters and requirements requires a broad knowledge of various emerging wireline and wireless communications and networking technologies.

Furthermore, to investigate a suitable information infrastructure, one needs to consider the available and emerging technologies, with potential to support the integration of smart microgrids. Formation of microgrids is becoming an increasingly attractive option to facilitate DER for their potential to increase the penetration rate of renewable resources. A microgrid may operate in one of two steady-state modes: grid-connected (i.e. connected to the main grid) and islanded (i.e. disconnected from the main grid). Transition from islanded mode to grid-connected mode involves re-synchronization. A smooth, well-managed, and controlled transient mode is critical to avoid negative impacts to the integrated grid as well as to the other grid-connected microgrids. Across the integration layer of grid-connected microgrids, information exchanges regarding measurement, monitoring, control, and management attributes need to be supported by various heterogeneous communications segments, including sensor networks, HANs, LANs,(local-area networks), MANs (metropolitan-area networks), WAN and global back-bone networks. On one hand, the information infrastructure to support integration of smart microgrids needs to share parts of public networks for global coverage. On the other hand, it requires a certain level of robustness sufficient to support reliable and timely delivery of vital information. For example, in islanded mode, the major part of the energy control and management can be confined within a particular microgrid. Accordingly, vital information can be exchanged within a private communications segment while information flowing in public communications segments is for routine reporting. Robust communications in a private segment can be locally and hence easily controlled. However, information flow in public communications segments may attempt to alert or request switching from an islanded mode to grid-connected mode, it will need a special treatment to get reliably delivered on time. This can be done by dynamically upgrading its quality-of-service class in order to obtain the required attention of the public segment.

In that regard, the capabilities, properties, performance, and standards of existing and emerging wireline and wireless communications technologies need to be reviewed to develop the most suitable information infrastructure that efficiently and effectively supports the integration of smart microgrids. Specifically, their provisions and limitations in various layers need to be examined, especially the physical layer (e.g. transmission capacity, quality, access protocols, operating frequency bands, environmental conditions), data layer (e.g. error protection and resilience, formats), and network layer (e.g. quality-of-service supports, networking configurations). In addition, the interworking and internetworking between candidate communications technologies and networks need to be considered to ensure the ability to configure a service-oriented infrastructure that efficiently governs exchange of data and commands between all components of a microgrid and within an integrated grid of smart microgrids. This means that various wireline and wireless communications segments need to be abstracted to facilitate the seamless device data integration into the information and automation descriptions and operation. Furthermore, communications, information processing, and networking techniques to enhance the performance of the information infrastructure need to be explored. Here, the focus should be on the development of performance-enhancing schemes that can be directly applied to the available and emerging communications systems, rather than techniques that require major changes. For example, instead of processing techniques at the signal level that involve redesigns of the physical layer, one needs to concentrate on processing at the bit level or higher. By exploring the hierarchical relationship among the nodes of the underlying communications infrastructure, cooperative transmission schemes can be developed that intelligently combine and process their information at the network level to improve transmission reliability and efficiency. One candidate approach is network coding by using combined coding at neighbor nodes. To avoid random and long delivery delay due to re-transmission in TCP/IP, forward-correction rateless coding is another potential scheme. Moreover, the knowledge of hierarchical information structures can be used to explore reconfigurable transmission and information processing strategies that enhance the communications robustness, leading to the development of quality-of-service parameters and establishment of dynamic quality-of-service requirements of different information types in smart microgrids, and the development of efficient transmission and information processing techniques and re-configurable, service-oriented networking architectures suitable for a robust information infrastructure to support the integration of smart microgrids and based on available and emerging communications technologies. To that end, one needs to characterize different information types in smart microgrids to establish their quality-of-service parameters and to classify their dynamic quality-of-service requirements. That will in turn drive the development of grid-integration requirements, and standards, codes and regulatory issues of emerging communications systems in supporting smart microgrids, as well as robust transmission techniques required for information exchange, based on jointly reconfigurable transmission and intelligent information processing schemes over heterogeneous wireless/wireline communications networks for integration of microgrids. That work would then need to be extended toward the development and evaluation of interworking and integration strategies for efficient transmission and information processing techniques and re-configurable, service-oriented networking architectures suitable for a robust information infrastructure to support the integration of smart microgrids and based on available and emerging communications technologies.

### 12.1.3.3 Topic 3.3: Distribution Automation Communications: Sensors, Condition Monitoring, and Fault Detection (Topic Leader: Meng; Collaborators: Chang, Li, Iravani, Farhangi, NB Power)

A robust and reliable sensor network is needed to provide information on overall grid integrity. As discussed earlier, microgrid communication infrastructure and the integrated data-management system are expected to facilitate the transfer of this information to central supervisory control. This is an important requirement for the intentional islanding operation to enable its ride-through capability when disconnected from the grid, and for the safety and power quality issues during unintentional islanding conditions. For re-connection, it is important for the microgrid to determine when the grid recovery occurs for grid re-synchronization.

The basic building block of an integrated sensor network for a smart grid is essentially a specialized communications and sensor module that is integrated and/or embedded within, ideally, all components of the smart microgrid to detect faults and measure the required system parameters such as power quality, voltage and frequency stability. This will require the development of hardware and firmware (in combination with analog sensors) that is capable of determining unhealthy grid operation in a fast and accurate manner as defined by the relevant standards requirement (e.g. currently in-draft IEEE Standard 1547.4). This information will be conveyed for distributed automation of the microgrid using the appropriate communication protocols for the purpose of islanding operation and protection.

*Objective I.* Develop technology-agnostic topology for Intelligent Sensor Network, including gateway access nodes with cellular capabilities that can provide an access point for sensor data.

*Objective II.* Cost-effective technologies for realization and integration of Intelligent Sensor Network. This should facilitate real-time processing of data, as well as to minimize transmission power and latency times that are required to ensure effective communication. Moreover, different communication scenarios, delays, and throughputs for sensory data need to be characterized.

*Objective III.* Optimal RTOS (Real Time Operating System) are needed to support dynamically changing sensor network profile. Efficient RTOS are of particular importance to this topic as the required performance from sensor network could be quite critical in certain areas of smart microgrids.

**Challenges** It is well understood that microgrids need to allow for multiple connection points between the microgrid and the main grid depending on the types of generation sources, penetration, and reliability requirements. This topology requires greater complexity in the control and protection of microgrid energy sources and loads, as well as the protection of the main grid. Microgrid monitoring is an important requirement for the intentional islanding operation of the microgrid and enables its ride-through capability when disconnected from the grid, and, for safety and power quality issues during unintentional islanding conditions. Also, for re-connection, it is important for the microgrid to determine when the grid recovery occurs for grid re-synchronization. A ride-through capability via intentional islanding control must be present to maintain voltage and frequency stability for the local microgrid loads. In this case, disconnecting from the main grid must be performed in a timely manner through a fault-detection process. Intentional islanding control must aggregate the capabilities of storage and energy sources to satisfy this requirement. Also, microgrid protection, through isolation, is generally required for common disturbances, such as lightning strikes, equipment malfunction, and downed power lines, on the main grid. Typically a smart microgrid will respond by tripping off-line until the grid recovers. As such, re-synchronization of the microgrid is required when a re-connection is requested. Synchronization (frequency and phase) is a global control concept that requires individual control and monitoring of all energy and storage devices within the microgrid such that satisfactory re-connection to the grid can be made. In the future, grid penetration from microgrid energy will be more prevalent. When this occurs, stable operation of microgrid as a net provider of energy will be of significant importance. Again, the operation of the microgrid energy sources and storage devices must work in a cohesive manner to deliver quality "dispatchable" power to the grid. Correspondingly, monitoring of each power source in the microgrid is required for an effective energy bidding/pricing strategy. Given these challenges, it is suggested that each microgrid energy source be enabled with fault detection and monitoring capability so that complete distributed control can be accomplished, and the challenges indicated above can be overcome.

Moreover, there are numerous challenges in the realization of power system monitoring and sensor hardware and firmware that are directly associated with a power grid connection. Previous fault detection methods can be grouped into active and passive schemes. Active schemes are more accurate in detecting grid faults but are more invasive due to energy injection into the grid. Passive schemes require less technology but are more susceptible to false-positive detections. In any event, the development of fault detection and grid-condition monitoring hardware and firmware is a difficult and tedious process given the nature of the voltage levels for grid-connected systems. The combination of digital processing and high voltage power electronic circuits requires extensive expertise, especially in the test and debug stages.

**Methodology**  Accurate condition monitoring of the grid should be implemented using state-of-the-art digital signal processor (DSP) controllers, as well as voltage and current sensor technologies suitable for microgrid applications. Pre-existing fault detection method could be selected based on a comprehensive literature search and through a simulation vetting process. The method should then be implemented and verified. Care must be taken in determining the latency requirements of the fault detection and monitoring system. For Supervisory Control and Data Acquisition (SCADA)-like requirements, a TCP/IP solution, based on best "effort," may not suffice and dedicated communication channels may be required, e.g. a wireless system with proprietary protocols. Moreover, grid monitoring technologies will have to focus on quantifying various passive and active grid-monitoring approaches and formulate a solution strategy that is cost-effective and satisfies the desired performance criteria.

### 12.1.3.4 Topic 3.4: Integrated Data Management and Portals

This topic focused on developing methodologies for managing massive amounts of real-time data, as well as designing temporally dynamic databases for storage of real-time data, and designing customer/utility information portals.

*Objective I.* Study the anatomies of highly versatile intelligent agents at various command and control layers within and across smart microgrids. Develop different command and control layers of a distributed management system for microgrid applications, requiring real-time data exchange within and across smart distribution networks and microgrids.

*Objective II.* Develop dynamically scalable multi-port database architecture to support local and remote energy management applications. This objective is particularly important for applications in which access to quasi-real-time AMI (Advanced Metering Infrastructure) data is required for realization of adaptive energy conservation and optimization solutions such as VVO (Volt-VAR Optimization).

*Objective III.* Platform-dependent architecture for user and utility portals with their associated presentation and visualization technologies. This requires a reliable real-time co-simulation platform that enables users and utilities to test different distribution automation applications using advanced communication standards/protocols such as DNP3 and IEC 61850. The architecture of this platform should be flexible enough to allow different DA applications to be tested with only a few changes in the settings.

**Challenge**  Introducing smart components into the utility distribution network places a huge burden on the utilities' communication infrastructure. Such components produce massive bursts of data (on polled or event basis) which need to be managed, understood, and stored. The nature of collected data is a function of its origin. Some data could indicate the imminence of alarms, while others would be for informational purposes only. Regardless of the nature of such data, its enormity and size make it impossible to manage through a simple process of collection, storage, and transference. Intelligence needs to be built into various interfaces to make sense of the data locally, aiding local decision making, and passing on to upstream devices only those pieces of information relevant to upstream functions. This is at the core of a distributed command and control system which is a hallmark of a smart grid, and a notable departure from the hierarchical command and control system adopted by legacy electrical grid.

*Methodology*  Intelligent agents are autonomous software entities, locked to a particular environment, which are capable of capturing information from other agents or from their surroundings to make independent decisions on the actions they need to perform. They work collaboratively with other agents in the same environment or elsewhere to achieve system objectives. They analyze the information, negotiate with other agents, take pre-emptive actions, and react in a timely fashion to the events in their environment. Intelligent agents could belong to an organization, such as infrastructure upkeep, looking after maintenance issues that would affect the entire system. They could also belong to a region, taking attributes and parameters set for that region, such as pricing tariffs and rebates. They could also be tasked with specific functions in a particular environment, such as alarm/status reporting. All such attributes will determine the extent of their functionality and operational scope. Smart grids are composed of independent processors that work together to deliver a system's functionality by supporting multitudes of intelligent agents that create an enormous amount of real-time data. Managing this data will enable intelligent agents within all data-producing/data-consuming components and terminations in a smart microgrid. Innovative anatomies and structures for embedded intelligent agents, real-time databases, and portals have not been tried out elsewhere. However, the fact that the concepts for such technologies have been around and they have been implemented, albeit in other forms elsewhere, is a strong indication of the feasibility of this approach. The development of highly efficient agent/data-base/portal structures that are capable of being cost-effectively implemented within smart components will be a major breakthrough.

Furthermore, one needs to find the most innovative topology for distributed command and control systems for a smart grid. The basic constituent components of such a control system are independent intelligent agents with innovative anatomies, capable of localized data processing and decision-making capabilities working collaboratively within the confines of dynamically changing global attributes. At the core of this research is the need to find structures for intelligent agents suitable for various components within the smart grid. Moreover, database architectures have to be developed that could be temporally dynamic and capable of virtualization. Innovative portal designs will have to be created to facilitate the evolution of information portals aimed at customer education to information portals allowing customers to act as stakeholders in energy transactions. The other topics in this theme deal with communication infrastructure, sensors, and standards. This topic focuses on ways for capturing and managing the data from termination points and sensors throughout the microgrid.

## 12.2   Final Thoughts

The Natural Sciences and Engineering Research Council Smart Microgrid Network (NSMG-Net) brought together some of best researchers and resources in what was previously the relatively unknown smart grid and microgrid areas to address various impediments that have slowed down or prevented implementation of smart microgrids in the world, including a lack of standards, universal regulatory frameworks, and robust technologies tested in near-real environments. Achieving the overarching goals of NSMG-Net required strong understanding and experience with all three disciplines at the heart of smart grids: *Power Systems, Communication Technology* and *Information Technology*. The multifaceted nature of the issues, and the need to leverage the expertise of a multi-sector group of experts, required a coordinated approach that the strategic network model provided very effectively. The network model helped create an environment in which researchers with divergent backgrounds, and no history of collaborative work, could find a common vocabulary to communicate, exchange information, and find interdisciplinary solutions for the multidimensional problems facing the utility industry. NSGM-Net researchers were able to solicit help, ideas and recommendations from experts from across the country to accelerate the development of innovative ideas that could be cost-effectively developed, implemented, and qualified.

The history of smart grid developments in academia and in the industry demonstrates the fact that the smart grid discipline does not replace, or supersede Power System Engineering. smart grid's main focus is on devising pervasive end-to-end data centric command and control technologies for the electricity grid. Such next-generation command and control technologies help raise the electricity grid's reliability, improve its efficiency, and reduce its carbon footprint (e.g. through integration of renewable sources of energy). As such, a smart grid should not be regarded as a replacement, or derivative or even next generation power system engineering. While traditional power system engineering deals with a wide variety of technological and architectural issues in power generation, transmission systems, and distribution networks, smart grid's focus is on (initially) the enhancement, and (eventually) replacement of electricity grid's overall command and control system, including architecture, technologies, and algorithms. In other words, as a discipline, smart grid simply complements traditional power system's command and control architecture and methodologies. It leverages communication systems and information technologies to create a new interdisciplinary technology platform, in which communication system engineers, information technology engineers, and power system engineers collaborate in the multifaceted multidisciplinary smart grid domain to facilitate the transition of the current electricity grid toward its future smart grid. It goes without saying that smart grid's interdisciplinary platform is further enriched with contributions from social scientists, business analysts, computer scientists, etc.

# References

**1** M. Ross, C. Abbey, Y. Brissette, et al. "Photovoltaic inverter characterization testing on a physical distribution system," in *2012 IEEE Power and Energy Society General Meeting*, 2012, pp. 1-7.

**2** Maitra, A., Pratt, A., Hubert, T. et al. (2017). Microgrid controllers: expanding their role and evaluating their performance. *IEEE Power Energy Mag.* 15: 41–49.

**3** H. Farhangi, "NSERC Smart Microgrid Research Network – Annual Report," NSERC – CRSNG2016

**4** "NSERC Smart Microgrid Network." 2012.[Online]. Available: http://www.nserc-crsng.gc.ca/Business-Entreprise/How-Comment/Networks-Reseaux/NSMGNet-NSMGNet_eng.asp

**5** M. Ross, C. Abbey, Y. Brissette, et al. "Real-time Microgrid Control Validation on the Hydro-Quebec Distribution Test Line," presented at the *CIGRE*, Paris, 2014.

**6** K. Rudion, A. Orths, Z.A. Styczynski, et al. "Design of benchmark of medium voltage distribution network for investigation of DG integration," in *2006 IEEE Power Engineering Society General Meeting*, 2006, p. 6 pp.

**7** "Technical Brochure 575: Benchmark Systems for Network Integration of Renewable and Distributed Energy Resources," final Report of CIGRE Task Force C6.04.02, 2010.

**8** A. Haddadi, A. Yazdani, G. Joos, et al. "A generic load model for simulation studies of microgrids," in *2013 IEEE Power & Energy Society General Meeting*, 2013, pp. 1-5.

**9** Yazdani, A. and Iravani, R. (2010). *Voltage-Sourced Converters in Power Systems: Modeling, Control, and Applications*. Wiley.

**10** I.M. Syed and A. Yazdani, "Simple mathematical model of photovoltaic module for simulation in Matlab/Simulink," in *Electrical and Computer Engineering (CCECE), 2014 IEEE 27th Canadian Conference on*, 2014, pp. 1-6.

**11** Farhangi, H. (2010). The path of the smart grid. *IEEE Power and Energy Magazine* 8 (1): 18–28.

**12** Etemadi, A.H., Davison, E.J., and Iravani, R. (2012). A decentralized robust control strategy for multi-DER microgrids – Part I: fundamental concepts. *IEEE Trans. Power Delivery* 27: 1843–1853.

**13** Etemadi, A.H., Davison, E.J., and Iravani, R. (2012). A decentralized robust control strategy for multi-DER microgrids – Part II: performance evaluation. *IEEE Trans. Power Delivery* 27: 1854–1861.

*Microgrid Planning and Design: A Concise Guide*, First Edition. Hassan Farhangi and Geza Joos.
© 2019 John Wiley & Sons Ltd. Published 2019 by John Wiley & Sons Ltd.

14 Tremblay, O. and Dessaint, L.-A. (2009). Experimental validation of a battery dynamic model for EV applications. *World Electr. Veh. J.* 3: 1–10.

15 Hansen, A.D. (2012). Generators and power electronics for wind turbines. In: *Wind Power in Power Systems* (ed. T. Ackerman), 73–103. Oxford: John Wiley & Sons, Ltd.

16 Torquato, R., Shi, Q., Xu, W. et al. (2014). A Monte Carlo simulation platform for studying low voltage residential networks. *IEEE Trans. Smart Grid* 5: 2766–2776.

17 Kamh, M.Z. and Iravani, R. (2012). A sequence frame-based distributed slack bus model for energy management of active distribution networks. *IEEE Trans. Smart Grid* 3: 828–836.

18 Manbachi, M., Nasri, M., Shahabi, B. et al. (2014). Real-time adaptive VVO/CVR topology using multi-agent system and IEC 61850-based communication protocol. *IEEE Trans. Sustainable Energy* 5: 587–597.

19 M. Manbachi, H. Farhangi, A. Palizban, et al. "Predictive algorithm for Volt/VAR optimization of distribution networks using Neural Networks," in *2014 IEEE 27th Canadian Conference on Electrical and Computer Engineering (CCECE)*, 2014, pp. 1-7.

20 Manbachi, M., Farhangi, H., Palizban, A. et al. (2016). Maintenance scheduling of volt-VAR control assets in smart distribution networks using advanced metering infrastructure. *Can. J. Electr. Comput. Eng.* 39: 26–33.

21 Manbachi, M., Farhangi, H., Palizban, A. et al. (2016). A novel volt-VAR optimization engine for smart distribution networks utilizing vehicle to grid dispatch. *Int. J. Electr. Power Energy Syst.* 74: 238–251.

22 M. Manbachi, A. Sadu, H. Farhangi, et al. "Real-time co-simulated platform for novel Volt-VAR Optimization of smart distribution network using AMI data," in *Smart Energy Grid Engineering (SEGE), 2015 IEEE International Conference on*, 2015, pp. 1-7.

23 M. Manbachi, A. Sadu, H. Farhangi, et al. "Real-time communication platform for smart grid adaptive Volt-VAR Optimization of distribution networks," in *Smart Energy Grid Engineering (SEGE), 2015 IEEE International Conference on*, 2015, pp. 1-7.

24 Manbachi, M., Sadu, A., Farhangi, H. et al. (2016). Real-time co-simulation platform for smart grid volt-VAR optimization using IEC 61850. *IEEE Trans. Ind. Inf.* 12: 1392–1402.

25 M. Nasri, H. Farhangi, A. Palizban, et al. "Multi-agent control system for real-time adaptive VVO/CVR in Smart Substation," in *Electrical Power and Energy Conference (EPEC), 2012 IEEE*, 2012, pp. 1-7.

26 Sinha, S. and Chandel, S.S. (2014). Review of software tools for hybrid renewable energy systems. *Renewable Sustainable Energy Rev.* 32: 192–205.

27 A.F. Stephanie Hay, "A Review of Power System Modelling Platforms and Capabilities," The IET 2015.

28 Faruque, M.D.O., Strasser, T., Lauss, G. et al. (2015). Real-time simulation technologies for power systems design, testing, and analysis. *IEEE Power Energy Technol. Syst. J.* 2: 63–73.

29 Banerjee, B., Jayaweera, D., and Islam, S. (2016). Modelling and simulation of power systems. In: *Smart Power Systems and Renewable Energy System Integration* (ed. D. Jayaweera), 15–28. Cham: Springer International Publishing.

**30** Joos, G., Reilly, J., Bower, W. et al. (2017). The need for standardization: the benefits to the core functions of the microgrid control system. *IEEE Power Energy Mag.* 15: 32–40.

**31** "IEEE Standard for the Specification of Microgrid Controllers," IEEE *Std 2030.7-2017*, pp. 1-43, 2018.

**32** Zamani, M.A., Yazdani, A., and Sidhu, T.S. (2012). A control strategy for enhanced operation of inverter-based microgrids under transient disturbances and network faults. *IEEE Trans. Power Delivery* 27: 1737–1747.

**33** Haddadi, A., Yazdani, A., Joos, G. et al. (2014). A gain-scheduled decoupling control strategy for enhanced transient performance and stability of an islanded active distribution network. *IEEE Trans. Power Delivery* 29: 560–569.

**34** M.Z. Kamh, R. Iravani and T.H.M. EL-Fouly, "Realizing a smart microgrid — Pioneer Canadian experience," *2012 IEEE Power and Energy Society General Meeting*, San Diego, CA, 2012, pp. 1–8. Fig. 5.

**35** "IEEE Recommended Practice for Excitation System Models for Power System Stability Studies," *IEEE Std 421.5-2016 (Revision of IEEE Std 421.5-2005)*, pp. 1-207, 2016.

**36** Bloemink, J.M. and Iravani, M.R. (2012). Control of a multiple source microgrid with built-in islanding detection and current limiting. *IEEE Trans. Power Delivery* 27: 2122–2132.

**37** Ajaei, F.B., Farhangi, S., and Iravani, R. (2013). Fault current interruption by the dynamic voltage restorer. *IEEE Trans. Power Delivery* 28: 903–910.

**38** Yazdanpanahi, H., Li, Y.W., and Xu, W. (2012). A new control strategy to mitigate the impact of inverter-based DGs on protection system. *IEEE Trans. Smart Grid* 3: 1427–1436.

**39** Mehrizi-Sani, A. and Iravani, R. (2012). Online set point adjustment for trajectory shaping in microgrid applications. *IEEE Trans. Power Syst.* 27: 216–223.

**40** Mehrizi-Sani, A. and Iravani, R. (2012). Constrained potential function – based control of microgrids for improved dynamic performance. *IEEE Trans. Smart Grid* 3: 1885–1892.

**41** Ross, M., Abbey, C., Bouffard, F. et al. (2015). Multiobjective optimization dispatch for microgrids with a high penetration of renewable generation. *IEEE Trans. Sustainable Energy* 6: 1306–1314.

**42** A. Zamani, T. Sidhu, and A. Yazdani, "A strategy for protection coordination in radial distribution networks with distributed generators," in *IEEE PES General Meeting*, 2010, pp. 1-8.

**43** Zamani, M.A., Sidhu, T.S., and Yazdani, A. (2014). Investigations into the control and protection of an existing distribution network to operate as a microgrid: a case study. *IEEE Trans. Ind. Electron.* 61: 1904–1915.

**44** Zamani, M.A., Yazdani, A., and Sidhu, T.S. (2012). A communication-assisted protection strategy for inverter-based medium-voltage microgrids. *IEEE Trans. Smart Grid* 3: 2088–2099.

**45** Etemadi, A.H. and Iravani, R. (2013). Overcurrent and overload protection of directly voltage-controlled distributed resources in a microgrid. *IEEE Trans. Ind. Electron.* 60: 5629–5638.

**46** Grilo, A.P., Gao, P., Xu, W. et al. (2014). Load monitoring using distributed voltage sensors and current estimation algorithms. *IEEE Trans. Smart Grid* 5: 1920–1928.

**47** Davarpanah, M., Sanaye-Pasand, M., and Iravani, R. (2013). Performance enhancement of the transformer restricted earth fault relay. *IEEE Trans. Power Delivery* 28: 467–474.

**48** Ellis, A., Kazachkov, Y., Sanchez-Gasca, J. et al. (2012). A generic wind power plant model. In: *Wind Power in Power Systems*, (ed. T. Ackerman), 799–820. Oxford: John Wiley & Sons, Ltd.

**49** Wu, J., Rangan, S., and Zhang, H. (eds.) (2012). *Green Communications: Theoretical Fundamentals, Algorithms, and Applications*, Series in Plasma Physics, vol. 1. London: CRC Press ProQuest ebrary. Web. 9 March 2017. – Chapter 2.

**50** Herath, S.P., Tran, N.H., and Le-Ngoc, T. (2015). Optimal signaling scheme and capacity limit of PLC under Bernoulli-Gaussian impulsive noise. *IEEE Trans. Power Delivery* 30: 97–105.

**51** Ho, Q.D., Rajalingham, G., and Le-Ngoc, T. (2013). Geographic-based routing in smart grid's neighbor area networks. *Rev J. Electron. Commun.* 3: 6.

**52** Q.D. Ho, Y. Gao, G. Rajalingham, et al. "Robustness of the routing protocol for low-power and lossy networks (RPL) in smart grid's neighbor-area networks," in *2015 IEEE International Conference on Communications (ICC)*, 2015, pp. 826-831.

**53** Saleh, S.A., Aljankawey, A.S., Meng, R. et al. (2014). Antiislanding protection based on signatures extracted from the instantaneous apparent power. *IEEE Trans. Power Electron.* 29: 5872–5891.

**54** Saleh, S.A., Aljankawey, A.S., Meng, R. et al. (2016). Apparent power-based anti-islanding protection for distributed cogeneration systems. *IEEE Trans. Ind. Appl.* 52: 83–98.

**55** J. Jia and J. Meng, "Partial discharge impulsive noise in electricity substations and the impact on 2.4 GHz and 915 MHz ZigBee communications," in *2013 IEEE Power & Energy Society General Meeting*, 2013, pp. 1-5.

**56** J. Jia and J. Meng, "A novel approach for impulsive noise mitigation in ZigBee communication system," in *2014 Global Information Infrastructure and Networking Symposium (GIIS)*, 2014, pp. 1-3.

**57** Manbachi, M., Farhangi, H., Palizban, A. et al. (2015). Quasi real-time ZIP load modeling for conservation voltage reduction of smart distribution networks using disaggregated AMI data. *Sustainable Cities Soc.* 19: 1–10.

**58** Z. Guanchen, G. Wang, H. Farhangi, et al. "Residential electric load disaggregation for low-frequency utility applications," in *2015 IEEE Power & Energy Society General Meeting*, 2015, pp. 1-5.

**59** "Smart Grid Reference Architecture," CEN-CENELEC-ETSI Smart Grid Coordination Group 2012.

**60** M. Quashie and G. Joos, "A methodology to optimize benefits of microgrids," in *2013 IEEE Power & Energy Society General Meeting*, 2013, pp. 1-5.

**61** Farhangi, H. (ed.) (2016). *Smart Microgrids: Lessons from Campus Microgrid Design and Implementation*. Boca Raton: Taylor & Francis.

**62** Bhuiyan, F.A., Yazdani, A., and Primak, S.L. (2015). Optimal sizing approach for islanded microgrids. *IET Renewable Power Gener.* 9: 166–175.

**63** Juan Clavier MS Thesis "Integration of renewable energy and storage in remote communities: an economic assessment" McGill University, 2013.

**64** Farrokhabadi, M., Canizares, C.A., and Bhattacharya, K. (2017). Frequency control in isolated/islanded microgrids through voltage regulation. *IEEE Trans. Smart Grid* 8: 1185–1194.

**65** M. Ross, "Microgrid Control with a High Penetration of Renewable Generation," PhD, Department of Electrical and Computer Engineering, McGill University, 2015.

**66** Yazdanpanahi, H., Xu, W., and Li, Y.W. (2014). A novel fault current control scheme to reduce synchronous DG's impact on protection coordination. *IEEE Trans. Power Delivery* 29: 542–551.

**67** M. Chlela, G. Joos, M. Kassouf, et al. "Real-time testing platform for microgrid controllers against false data injection cybersecurity attacks," in *2016 IEEE Power and Energy Society General Meeting (PESGM)*, 2016, pp. 1-5.

**68** EPRI."Smart Grid Resource Center – Use Case Repository."[Online]. Available: http://smartgrid.epri.com/Repository/Repository.aspx

**69** Wang, Y., Yazdanpanahi, H., and Xu, W. (2013). *"Harmonic impact of LED lamps and PV panels," 2013 26th IEEE Canadian Conference on Electrical and Computer Engineering (CCECE)*, 1–4. SK: Regina.

**70** "IEEE Draft Standard for the Testing of Microgrid Controllers," *IEEE P2030.8/D12, March 2018*, pp. 1-43, 2018.

**71** Ross, M. and Quashie, M. (March 2016). NSMG-Net Project 2.6 Application of EMS and Valuation to BCIT Microgrid. *McGill University* .

**72** Martin, K.E. and Carroll, J.R. (2008). Phasing in the Technology. *IEEE Power Energy Mag* 6 (5): 24–33.

**73** Thorp, J.S., Abur, A., Begovic, M. et al. (Sept-Oct. 2008). Gaining a Wider Perspective. *IEEE Power Energy Mag.* 6 (5): 43–51.

# Index

*Microgrid Planning and Design: A Concise Guide*, First Edition. Hassan Farhangi and Geza Joos.
© 2019 John Wiley & Sons Ltd. Published 2019 by John Wiley & Sons Ltd.